Sustainable Finance

Sustainable Finance is a concise and authoritative reference series linking research and practice. It provides reliable concepts and research findings in the ever growing field of sustainable investing and finance, SDG economics and Leadership with the declared commitment to present the theories, methods, tools and investment approaches that can fulfil the United Nations Sustainable Development Goals and the Paris Agreement COP 21/22 alongside with de-risking assets and creating triple purpose solutions that ensure the parity of profit, people and planet through choice architecture passion and performance. The series addresses market failure, systemic risk and reinvents portfolio theory, portfolio engineering as well as behavioural finance, financial mediation, product innovation, shared values, community building, business strategy and innovation, exponential tech and creation of social capital. Sustainable Finance and SDG Economics series helps to understand keynotes on international guidelines, guiding accounting and accountability principles, prototyping new developments in triple bottom line investing, cost benefit analysis, integrated financial first plus impact first concepts and impact measurement. Going beyond adjacent fields (like accounting, marketing, strategy, risk management) it integrates the concept of psychology, innovation, exponential tech, choice architecture, alternative economics, blue economy shared values, professions of the future, leadership, human and community development, team culture, impact, quantitative and qualitative measurement, Harvard Negotiation, mediation and complementary currency design using exponential tech and ledger technology. Books in the series contain latest findings from research, concepts for implementation, as well as best practices and case studies for the finance industry. This book series is indexed in SCOPUS

Rachel Calipha • Dafna M. DiSegni
Shirit Katav-Herz
Editors

Sustainable Finance Regulation in the European Union

Pathways, Practices and National Perspectives

Springer

Editors
Rachel Calipha
Academic College of Tel Aviv-Yaffo
Tel Aviv-Yaffo, Israel

Dafna M. DiSegni
University of Rhode Island
Kingston, RI, USA

Shirit Katav-Herz
Academic College of Tel Aviv-Yaffo
Tel Aviv-Yaffo, Israel

ISSN 2522-8285 ISSN 2522-8293 (electronic)
Sustainable Finance
ISBN 978-3-031-88391-0 ISBN 978-3-031-88392-7 (eBook)
https://doi.org/10.1007/978-3-031-88392-7

This Springer imprint is published by the registered company Springer Nature Switzerland AG
The registered company address is: Gewerbestrasse 11, 6330 Cham, Switzerland

Preface

This book is edited by Califa, DiSegni, and Katav-Herz, with equal contribution from each editor.

The objective of the present book is to outline the background and transition to a sustainable financial system in the European Union. We provide insights into the key events and strategies that have shaped the EU's sustainable finance landscape, including the Paris Agreement and Action Plan 2018, as well as the renewed strategy and ESMA's roadmap. Furthermore, it elaborates on the regulations, tools, and standards encompassed by the sustainable finance framework, all of which are founded on this overarching strategy.

The review and analysis of the European Sustainable Finance framework is presented through six distinct perspectives, implemented by the following six nations: Spain, Italy, Estonia, Germany, the Netherlands, and Israel. These country-specific analyses showcase the practical implementation of sustainable finance regulations, with each nation dedicated to a chapter in Part II of the book. This section highlights the unique approaches and challenges faced by these countries in fostering sustainable financial practices. Our comparative analysis of the six national regulatory regimes offers a comprehensive understanding of the various frameworks, allowing readers to grasp the similarities and differences in sustainable finance approaches across the EU.

The book concludes with forward-looking recommendations for future policy implementation and addresses potential criticisms and obstacles that may arise in the journey toward a sustainable financial future. We consider this publication a valuable resource for policymakers, researchers, and industry practitioners, providing an in-depth exploration of sustainable finance regulation in the European Union and its implications for a greener and more sustainable world.

The collaborating authors have contributed to the presented themes, both through individual efforts and collaborative endeavors. The Inclusive Finance Initiative (IFI) project, fostered and financially supported by ERASMUS EU funds, has been instrumental in facilitating the collaborative synergy among the authors. This platform has not only nurtured the development of diverse themes but has also fostered the cultivation, exchange, and dissemination of ideas within the robust regulatory framework of sustainable finance.

Tel Aviv, Israel Rachel Calipha
Kingston, RI, USA Dafna M. DiSegni
Tel Aviv, Israel Shirit Katav-Herz

Acknowledgments

It is an undeniable fact that caring for our planet and society is a profound responsibility that falls on all of us as global citizens. The emergence of sustainable finance as an essential tool is a beacon of hope where capital flows are aligned with sustainability-driven efforts to mitigate the adverse impacts of climate change, reverse environmental degradation, promote social equity, and foster inclusive growth. The stakes could not be higher, and the urgency could not be greater. The future of our planet and society depends on our ability to coordinate our efforts and work together toward common goals. The transition to sustainable finance is not only a moral imperative but also an economic and social necessity. It is a call for action that requires a fundamental shift, not just in the financial sector but across various fields such as economics, technology, law, and society. It is indispensable to steer finance toward long-term resilience and multi-sector coordination through policy and regulation. I had the privilege to coordinate the Erasmus+ Capacity Building in Higher Education Project on *Innovative Finance Inclusion in Academic and Field (IFI)*[1] during the years 2021–2023. With a professional team from thirteen academic institutions in Israel and Europe, we joined forces to align the Israeli higher education system with global changes and develop new capacities for inclusive social finance on three levels: (a) ecosystem and infrastructure; (b) academy-industry cooperation; and (c) research, teaching, and training.

I am pleased to introduce this book, which is one of the valuable fruits of that collaborative endeavor. The book explores various aspects of sustainable finance regulation, starting with a review of its evolution in Europe throughout the years, continuing with an array of case studies that provide an in-depth analysis of implementing sustainable finance in six different countries, and finalizing with an integrative outlook toward the future. This book comes at a critical juncture, as the world is experiencing dramatic environmental, social, and economic changes in a turbulent era. The continuous efforts to build a sustainable finance framework through tools, standards, and regulations aimed at navigating the financial sector toward

[1] 619453-EPP-1-2020-1-IL-EPPKA2-CBHE-JP.

sustainability requires a fundamental transition—not just in finance but across disciplines. With the IFI project, we have embraced this interdisciplinary approach to sustainable finance and brought together partners from higher education institutions along with industry and government. It is our hope that this book will inspire policymakers, regulators, and industry practitioners to embrace the challenge of transitioning to a sustainable financial system. As the coordinator of the Erasmus+ IFI Project, I extend my appreciation to the editors and congratulate the collaborating authors, who have produced a significant contribution to the literature through their individual and joint efforts. I hope this book helps to disseminate ideas and best practices related to inclusive and sustainable finance, further strengthening the emerging ecosystem in this critical domain.

Vered Holzmann

The Academic College of Tel Aviv Yaffo, Israel

Contents

Editors and Contributors

About the Editors

Rachel Calipha is a Senior Lecturer at the Academic College of Tel Aviv Yaffo, Israel, where she heads the Capital Market specialty track and the Research Institute of Society and Economics (RISE Impact). Her research fields are strategy and finance.

Dafna M. DiSegni is a Visiting Professor of Economics at the University of Rhode Island. Her research focuses on natural resource markets, industrial organization, and the development of financial mechanisms for sustainability. Her research includes risk analysis and the valuation of the potential impacts of financial regulations on industry value chains, resource conservation, the development of backstop technologies, and the diffusion of knowledge, information, and sustainability. At the time of ERASMUS project, she was affiliated with the Tel Hai College, Israel.

Shirit Katav-Herz is a Senior Lecturer at the Academic College of Tel Aviv Yaffo, Israel, where she heads the Undergrad Program in Economics and Management. Her research fields are behavioral economics and public economics, focusing on the effect of social norms on consumer's decision-making and the implication on policy.

Contributors

Rachel Calipha Academic College of Tel-Aviv Yaffo, Tel-Aviv Yaffo, Israel

Nathan de Arriba-Sellier Rotterdam School of Management, Erasmus University, Rotterdam, Netherlands

Dafna M. DiSegni University of Rhode Island, Kingston, RI, USA

Estibaliz Goicoechea Deusto Business School (DBS), University of Deusto, Bilbao, Spain

Saadat Hussain Ruprecht-Karls-Universität Heidelberg, Heidelberg, Germany

Shirit Katav-Herz Academic College of Tel-Aviv Yaffo, Tel-Aviv Yaffo, Israel

Aleksandra Kekkonen Estonian Business School, Tallinn, Estonia

Kristiina Kerge Estonian University of Life Sciences, Tartu, Estonia

Mikel Larreina Deusto Business School (DBS), University of Deusto, Bilbao, Spain

Nadav Paldi Ministry of the Environment, Jerusalem, Israel

Patrizia Riva Università del Piemonte Orientale, Novara, Italy

Dirk Schoenmaker Rotterdam School of Management, Erasmus University, Rotterdam, Netherlands

Gal Tamir Israel National Digital Agency, Strategy and Policy Unit, Tel Aviv, Israel

Abbreviations

ABI	Associazione Bancaria Italiana
AEX 25	Amsterdam Exchange Index 25
AFM	Autoriteit Financiële Markten (Dutch Authority for the Financial Market)
AI	Artificial Intelligence
AMCESFI	Autoridad Macroprudencial Consejo de Estabilidad Financiera (Spanish Macroprudential Authority)
AMF	Autorité' des Marchés Financiers (French Authority for the Financial Market)
BaFin	Bundesanstalt für Finanzdienstleistungsaufsicht (Federal Financial Supervisory Authority)
BME	Bolsas y Mercados Españoles (Spanish Stock Exchanges and Markets)
BoSA	Board of Statutory Auditors
BSc	Bachelor of Science degree
BTAR	Banking Book Taxonomy Alignment Ratio
CAC 40	Cotation Assistée en Continu 40 (Paris Stock Market Index 40)
CIS	Collective Investment Schemes
CNMV	Comisión Nacional del Mercado de Valores (National Securities Market Commission)
CONSOB	Commissione Nazionale per le Società e la Borsa
COP21	21st Conference of the Parties
COVID-19	Coronavirus Disease 2019
CSDD	Corporate Due Diligence Directive
CSRD	Corporate Sustainability Reporting Directive
DAX 30	Deutscher Aktienindex 30 (German Stock Market Index 30)
DNB	De Nederlandsche Bank (Dutch Central Bank)
DNSH	Doing No Significant Harm
EBA	European Banking Authority
EC	European Commission
ECB	European Central Bank

EFRAG	European Financial Reporting Advisory Group
EIBIS	The European Investment Bank's Investment Survey
EIOPA	European Insurance and Occupational Pensions Authority
EsBA	The Estonian Banking Association
ESG	Environmental, Social, and Governance
ESMA	European Securities and Markets Authority
ESRB	European Systemic Risk Board
ESRS	European Sustainability Reporting Standards
EU	European Union
EUGBS	European Green Bond Standard
FLESB	Forward Looking Exercise on Spanish Banks
FNG	Forum Nachhaltige Geldanlagen
FTSE MIB 40	Financial Times Stock Exchange Milano Indice di Borsa 40 (Milan Stock Market Index 40)
GAR	Green Asset Ratio
GDP	Gross Domestic Product
GHG	Greenhouse gas
GRI	Global Reporting Initiative
GSIA	Global Sustainable Investment Alliance
GTAI	Germany Trade and Invest
IBEX 35	Iberia Index 35
ICAC	Instituto de Contabilidad y Auditoría de Cuentas (Institute of Accounting and Auditing)
ICJCE	Instituto de Censores Jurados de Cuentas de España (Institute of Chartered Accountants of Spain)
ICT	Information and communications technology
ID	Identity Document
IDD	Insurance Distribution Directive
ILO	International Labor Organization
ISA	International Standard on Assurance Engagements
ISD	Investment Services Directive
ISSB	International Sustainability Standard Board
KfW	Kreditanstalt für Wiederaufbau (Credit Institution for Reconstruction)
km^2	Square kilometer
KPIs	Key Performance Indicators
KRIs	Key Risk Indicators
kt	1000 ton
LSIs	Less Significant Institutions
LULUCF	Land use, land-use change, and forestry
MiFID II	Markets in Financial Instruments Directive II
MiFIR	Markets in Financial Instruments Regulation
MSc	Master of Science degree
NFGS	Network for Greening the Financial System
NFI	Non-Financial Information

NFRD	Non-Financial Reporting Directive
NFS	Non-Financial Statement
NZBA	Net-Zero Banking Alliance
PABs	Paris-aligned Benchmarks
PAIs	Principal Adverse Impacts
PNFS	Plan Nacional de Finanzas Sostenibles (National Plan for Sustainable Finance)
PPPs	Public–Private Partnerships
PRI	Principles for Responsible Investment
PSI 20	Portuguese Stock Index 20
R&D	Research and Development
RAF	Risk Appetite Framework
RD	Real Decreto (Royal Decree)
RDL	Real Decreto Ley (Royal Decree-Law)
REA	Registro Economistas Auditores (Registry of Auditing Economists)
RegTech	Regulatory Technology
RRF	Recovery and Resilience Facility
SDGs	Sustainable Development Goals
SF	Sustainable Finance
SFB	Sustainable Finance Beirat (Sustainable Finance Advisory Committee)
SFDR	Sustainable Finance Disclosures Regulation
SMEs	Small- and Medium-sized Enterprises
SupTech	Supervisory Technology
TCFD	Task Force Climate-related Financial Disclosures
TSI	Technical Support Instrument
UNESCO	United Nations Educational, Scientific and Cultural Organization
UNGC	United Nations Global Compact
US	United States

List of Figures

List of Tables

Part I
Foundations of Sustainable Finance Regulation

Chapter 1
Introduction

Dafna M. DiSegni

As public awareness of sustainability issues grows and the critical role of finance in driving sustainable development becomes increasingly clear, understanding the mechanisms behind sustainable finance is more relevant than ever. Governments, corporations, and individuals are recognizing that financial practices must evolve to address environmental challenges and promote long-term economic sustainability. This book offers a comprehensive look at sustainable finance within the European Union, exploring the detailed regulations and practices that guide its shift toward a more sustainable economy. It unfold the EU's complex sustainable finance framework, presenting the key strategies, regulatory achievements, and important agreements that define its progress. The book is divided into three carefully structured parts, each contributing to the intricate tapestry of regulations and practices that are steering the transition towards a greener economy.

Part I exposes the reader to the evolving landscape of sustainable finance regulations within the European Union (EU), exploring the multifaceted framework established over time to facilitate the realization of the EU's sustainable finance strategy. The intricate web of EU Sustainable Finance regulations encompasses several pivotal components: the EU taxonomy regulation and the EU disclosure regulations (Sustainable Finance Disclosure Regulation (SFDR)); Corporate Sustainability Reporting Directive (CSRD); Markets in Financial Instruments Directive II (MiFID); and Insurance Distribution Directive (IDD). These regulations wield authority across the spectrum of investment stakeholders, encompassing corporations seeking capital and investors supplying it, as well as all interstitial actors. Segueing further, the chapter navigates to the realm of the Disclosure Regulation. This segment delineates the distinct cohorts targeted by each regulation and related stakeholders: The issuers (CSRD), the market participants, and the financial

D. M. DiSegni (✉)
University of Rhode Island, Kingston, RI, USA
e-mail: dafna.disegni@uri.edu

R. Calipha et al. (eds.), *Sustainable Finance Regulation in the European Union*, Sustainable Finance, https://doi.org/10.1007/978-3-031-88392-7_1

advisers (SFDR; MiFID II and IDD). We wrap up this first part by exploring the EU's pivotal tools and standards, including a description of the EU Climate Transition Benchmarks, EU Paris-aligned Benchmarks, and the European green bond standard.

In Part II (Chaps. 4–9), we provide a comprehensive analysis of six distinct countries' endeavors to adapt and implement a financial framework in alignment with the European Union's sustainable finance directives. We explore of diverse national approaches to sustainable finance regulations, highlighting the unique strategies each country has employed. The countries under scrutiny encompass Spain, Italy, Germany, Estonia, the Netherlands and Israel, each contributing a distinctive perspective to the global sustainable finance landscape.

Chapter 4 details Spain's unique stance towards sustainable finance. There, much attention is directed to the financial support to environmental sustainability, fostering transition to alternative energy vehicles, building efficiency standards, and legislation for a circular economy and environmental responsibility. Social sustainability is underpinned by this framework promoting equality, non-discrimination, and equitable work conditions, exemplifying Spain's comprehensive integration of ESG principles in finance.

In Chap. 5, the Italian Paradigm is discussed. This chapter delves into Italy's approach, offering a glimpse into the local adaptations aligned with EU objectives. Italy's approach to sustainable finance regulations is characterized by a comprehensive and integrated framework, focusing on environmental, social, and governance (ESG) aspects. Key features include (1) Constitutional and Legislative Reforms: Italy's legal tradition emphasizes ESG themes stemming from its 1948 Constitution. Recent reforms, such as Constitutional Law 1/2022, emphasize protecting the environment and public health, aligning private economic activities with these goals. (2) National Action Plans and Strategies: The Italian National Action Plan on Business and Human Rights, the National Transition Plan for Ecological Transition, and the National Strategy for Circular Economy demonstrate Italy's commitment to sustainable development goals and transitioning to a green economy. (3) National Recovery and Resilience Plan (PNRR): This includes investments and reforms in health, environment, biodiversity, and climate, promoting a circular economy and renewable energy sources. (4) Legislative Decree 152/2006 on environmental liability, laws mandating the recruitment of disabled workers, anti-corruption models for state-owned companies, and the Sustainable Finance Disclosure Regulation (SFDR), which establishes a framework for sustainable investment applicable to Italian market operators. (5) Green Procurement Laws: Law 221/2015 and related decrees promote green economy measures and establish minimum environmental criteria in public procurement, ensuring environmentally responsible purchasing practices. These regulations collectively form a robust and multi-dimensional approach to sustainable finance, reflecting Italy's commitment to integrating ESG principles into its economic and social fabric.

Chapter 6 focuses on Germany, shedding light on its contributions to sustainable finance. Germany's sustainable finance strategy, as outlined by BaFin (the financial regulatory authority for Germany), hinges on five key factors: (1) Risk-Oriented

Regulation: BaFin endorses a practical approach to regulation, stressing that green investments should be risk-assessed without inherent assumptions of reduced risk; (2) Reliable Data: Acknowledging the need for quality data, BaFin anticipates improvements with the EU's Corporate Sustainability Reporting Directive (CSRD) and expects to enhance enforcement capabilities for compliance; (3) Management of Environmental Risks: BaFin urges market participants to better recognize and manage environmental risks, ensuring transparent disclosure of the transition to sustainability and physical risks; (4) Combating Greenwashing: To prevent misleading claims about sustainability, BaFin focuses on ensuring clear disclosure and honesty in sustainability profiles and investment products; (5) Knowledge Sharing: BaFin emphasizes the value of dialogue with stakeholders and experts and contributes to ESG discussions on a national and international level.

Chapter 7 describes Estonia's distinctive approach to sustainable finance regulations, offering insights into the nation's tailored implementation of EU guidelines. Estonia is currently at the stage of finalizing its SF roadmap towards sustainability. The agricultural sector is presented as a central economic activity that gains attention concerning the sustainable finance framework to support it. This sector's significance is due to its impact on the economy and the environment, making it a priority to apply sustainable practices funded through innovative financial mechanisms that align with international sustainability commitments.

Chapter 8 extends the discussion to the Netherlands' approach. New EU transparency and conduct of business regulations have reshaped the Dutch Authority of Finance. While sustainability was not a core part of AFM's strategy at the onset of the 2020s, it has become one of its priorities, primarily due to rising investor demand in a country exposed to unique climate risks and hosting a robust financial sector. This chapter describes how the Netherlands built its capacity via SF regulations. Its early focus has been surveying market trends and guiding the market, mainly due to the implementation challenges of SFDR and the CSRD. The AFM has also sought to weigh in on the development of EU regulation, often jointly with the French regulator. Increasingly, the AFM is turning towards enforcement, notably to prevent greenwashing and risk management, as physical and transition risks become more evident.

We conclude Part II by implementing the Sustainable Finance strategy in Israel, in Chap. 9. Incorporating Israel as an additional case study illuminates the spillover of EU regulatory approaches on non-EU countries. Israel has been increasingly proactive in advancing sustainable finance initiatives, driven by a growing awareness of environmental and social challenges and a desire to position itself in the front of innovation and responsible finance. The local Capital Market, Insurance, and Savings Authority (CMISA) has recommended integrating ESG factors into investment and risk management since 2020. Sustainability reporting is not mandatory; however, the Israeli Institute of Certified Public Accountants (ICPA) has released guidelines for voluntary ESG disclosures. The ISA urges climate-related disclosures, and the CMISA emphasizes including ESG factors in corporate governance. Overall, there's a growing recognition of the financial implications of climate change and sustainability, promoting awareness among investors regarding the associated risks and opportunities.

Part III (Chaps. 10 and 11) of the book undertakes a systematic cross-country analysis, evaluating the regulatory paradigms embraced by the selected nations about crucial dimensions, such as the extent of adoption and implementation, the significance of social inclusion within sustainable finance regulations, similarities in enforcement and compliance challenges. To this end, Chap. 10 focuses on the broader social dimensions, emphasizing the significance of social inclusion and communities' engagement in driving sustainable finance forward. It explores how involving diverse stakeholders can amplify positive outcomes and equitable benefits. At last, Chap. 11 provides a compelling synthesis of the journey undertaken throughout the book. It distills the main findings and insights discussed in previous chapters, culminating in a comprehensive overview of the sustainable finance landscape. Through a cohesive narrative, this chapter reaffirms the pivotal role of sustainable finance regulation within the unique contexts of the countries examined and its broader alignment with the EU framework. This chapter offers a reflective vantage point by threading the intricate threads of regulatory frameworks, cross-country comparisons, challenges, opportunities, and the trajectory of sustainable finance. It reinforces the imperative for ongoing collaboration, innovation, and adaptation to steer towards a future that harmonizes sustainability goals with financial resilience, echoing the resounding significance of this collective journey in advancing sustainable finance in both national and international arenas.

Join us as we navigate the complexities of sustainable finance regulation. We offer an insightful and dynamic narrative designed to spark transformative change for a sustainable world.

Dafna M. DiSegni is a Visiting Professor of Economics at the University of Rhode Island. Her research focuses on natural resource markets, industrial organization and the development of financial mechanisms for sustainability. Her research includes includes risk analysis and the valuation of the potential impacts of financial regulations on industry value chains, resource conservation, the development of backstop technologies, and the diffusion of knowledge, information, and sustainability. At the time of ERASMUS project she was affiliated with the Tel Hai College, Israel.

Chapter 2
The European Union Sustainable Finance Strategy

Shirit Katav-Herz and Rachel Calipha

Abbreviations

EC	European Commission
ESG	Environmental, Social, and Government
ESMA	European Securities and Markets Authority
SDGs	Sustainable Development Goals
SF	Sustainable Finance

2.1 Introduction

Sustainability is crucial for a safer and better world for all. Multiple definitions of sustainability emerge from different disciplines. In ecology, for example, sustainability relates to how biological systems remain productive over time (Bromley, 2008). Since the 1980s, the term has increasingly been associated with the concept of human sustainability on planet Earth, giving rise to the most widely quoted definition of sustainable development, which is defined by the UN's Brundtland Commission (March 20, 1987) as follows: "Sustainable development is a development that meets the needs of the present without compromising the ability of future generations to meet their own needs."

Sustainable development encompasses environmental (e.g., natural resources use, pollution prevention, and biodiversity), economic (e.g., growth, profit, cost saving, and R&D), and social (e.g.standard of living, education, jobs, and equal opportunity) aspects. It also includes the interplay between and among these three aspects, such as social-economic aspects (e.g., business ethics, fair trade, and workers' benefits), environmental, economic aspects (e.g., energy efficiency, renewable fuels,

S. Katav-Herz (✉) · R. Calipha
Academic College of Tel Aviv-Yaffo, Tel Aviv-Yaffo, Israel
e-mail: Shirit@mta.ac.il; Rachelca@mta.ac.il

R. Calipha et al. (eds.), *Sustainable Finance Regulation in the European Union*, Sustainable Finance, https://doi.org/10.1007/978-3-031-88392-7_2

subsidies, and incentives), and social-ecological elements (e.g., conservation policies, environmental justice, and global stewardship) (Rodriguez et al., 2002).

The European Union is committed to sustainable development that meets the needs of present and future generations while opening up new employment and investment opportunities and ensuring economic growth. The EU entrusts the European Commission with the task to carry out its commitment effectively. The European Commission serves the broader interests of the EU by formulating and enforcing legislation and implementing policies and the EU budget. As the EU's politically independent executive arm, the European Commission assumes sole responsibility for drafting new proposals for European legislation and implementing the directives of the European Parliament and the Council of the EU (EC, n.d.). Two key events accelerated the adaptation of sustainable development: the 2030 Agenda for Sustainable Development and the Paris Agreement. They are briefly discussed below:

- **The 2030 Agenda for Sustainable Development**

  On 25 September 2015, the UN General Assembly adopted newglobe sustainable development framework: The 2030 Agenda for Sustainable Development (the '2030 Agenda'). The 2030 Agenda has the Sustainable Development Goals (SDGs) at its core and covers the three dimensions of sustainability: Economic, social, and environmental. The Commission's communication of November 22, 2016, on the "next steps for a sustainable European future" links the SDGs to the Union's policy framework to ensure that all EU actions and policy initiatives, both within the Union and globally, take the SDGs on board at the outset. In its conclusions of June 20, 2017, the Council confirmed the commitment of the Union and its Member States to the implementation of the 2030 Agenda in a full, coherent, comprehensive, integrated, and effective manner, in close cooperation with partners and other stakeholders (UN, n.d.-a, n.d.-b).

- **The Paris Agreement**

  The Paris Agreement, ratified in December 2015 by 195 nations, marks an unprecedented global accord aimed at enhancing adaptation, fostering resilience to climate change, and capping global temperature rise to significantly below 2 °C of business as usual.

  The Union approved the Paris Agreement adopted under the UN Framework Convention on Climate Change on October 5, 2016. Article 2(1)(c) of the Paris Agreement aims to strengthen the response to climate change by making finance flows consistent with a pathway towards low greenhouse gas emissions and climate-resilient development, among other means. In this context, on December 12, 2019, the European Council adopted conclusions on climate change. In light of this, the EU climate law has entered into force and represents a crucial step towards achieving a climate-neutral Union by 2050. By adopting the Paris Agreement on climate change and the UN 2030 Agenda for Sustainable Development in 2015, governments worldwide chose a more sustainable path for our planet and economy. At the core of the UN 2030 Agenda are 17 Sustainable Development Goals (SDGs). Over the next 15 years, these goals will guide the world in preparing for a future that supports stability, a healthy planet, fair, inclusive, resilient societies, and prosperous economies (EUR-Lex, 2020).

The financial system's involvement is essential to realizing these goals. This chapter explores the financial system's pivotal role and elucidates the concept of sustainable finance. Subsequently, it unveils the EU's sustainable finance strategy, beginning with the initial action plan and tracing its evolution through subsequent updates. Ultimately, the chapter outlines the critical role of the European Securities and Markets Authority (ESMA) and its strategic approach to fostering sustainable finance.

2.2 The Role of the Financial System in Sustainable Development

As we are increasingly faced with the risks and consequences of climate change and resource depletion, urgent action is needed to adapt public policies to this new reality. The financial system has a crucial role to play here. The financial system is being reformed to address the lessons of the financial crisis. In this context, it can be part of the solution toward a greener and more sustainable economy. Reorienting private capital to more sustainable investments requires a comprehensive shift in how the financial system works. This shift is necessary if the EU is to develop more sustainable economic growth, ensure the financial system's stability, and foster more transparency and long-termism in the economy (EC, 2017). The financial system supports the economy by funding economic activities, jobs, and growth. Investment decisions are typically based on several factors. Still, those related to environmental and social considerations are often not sufficiently considered since such risks are likely to materialize over a longer period. It is essential to recognize that taking longer-term sustainability interests into account makes economic sense and does not necessarily lead to lower returns for investors (DiSegni et al., 2015). Here is what key figures have said in this context:

Europe can lead by the power of its example. Europe has become the home of sustainable investment. We are proud of this leadership, and we will keep consolidating." Ursula Vonder Leyen, 13th president of the European Commission, 2021.

The sustainability transition requires a collective effort of several parties. Our role as regulators and supervisors is to ensure that a sound regulatory regime and that timely action is taken when there is a threat to investment or protection or the overall stability of the financial system." Verena Ross, Chair of the European Securities and Markets Authority (ESMA), November 2021.

2.3 European Sustainable Finance Strategy

'Sustainable finance' generally refers to considering environmental and social considerations in investment decision-making, leading to increased investments in longer-term and sustainable activities. More specifically, ecological considerations refer to climate change mitigation, adaptation, and the environment more broadly and related risks (e.g., natural disasters). Social considerations may refer to inequality, inclusiveness, labor relations, investment in human capital, and communities. Environmental and social concerns are often intertwined, as climate change can significantly exacerbate existing inequality. The governance of public and private institutions, including management structures, employee relations, and executive remuneration, plays a fundamental role in ensuring the inclusion of social and environmental considerations in the decision making process (EC, n.d.).

In late 2016, the European Commission established a High-Level Expert Group on sustainable finance. By January 31, 2018, the expert group published its final report, presenting an exhaustive roadmap for developing a sustainable finance strategy within the European Union. The report identifies two critical drivers underpinning sustainable finance: firstly, enhancing the financial sector's role in supporting sustainable and inclusive growth by addressing society's long-term needs, and secondly, bolstering financial stability through integrating Environmental, Social, and Governance (ESG) considerations into investment decisions. This report features eight principal recommendations, encompassing a variety of comprehensive measures along with actions tailored specifically for different sectors within the financial system. This recommendation became part of the action plan, which was further discussed.

On March 8, 2018, the European Commission adopted its Action Plan on Financing Sustainable Growth, proposing concrete actions to help achieve the SDGs and objectives outlined in the Paris Agreement on Climate Change. In this Action Plan (see details in the list below), the European Commission has called on other players, such as EU member states, to support the implementation of sustainable finance actions and promote this transformation in their territories. The European Commission is creating a dedicated EU member states expert group on sustainable finance to facilitate this. The Action Plan aims to: Reorient capital flows towards sustainable investment to achieve sustainable and inclusive growth, reduce the environmental footprint on the planet, and boost competitiveness by improving the efficiency of production processes and reducing the costs of accessing and managing resources. Manage financial risks stemming from climate change, resource depletion, environmental degradation, and social issues by including environmental and social goals in financial decision-making towards limiting the financial impact of environmental and social risks. Foster transparency and long-term financial and economic activity. Increased transparency, underpinned by innovative technologies, empowers citizens to compare the sustainability performance of companies and allows retail investors to make informed investment decisions. Corporate transparency on sustainability issues is a prerequisite for enabling financial market actors to properly assess companies' long-term value creation and manage sustainability risks.

List of the ten actions:

Action 1: Establishing an EU classification system for sustainable activities. The aim is to embed the future EU sustainability taxonomy in EU law and provide the basis for using such a classification system in different areas (e.g., standards, labels, green-supporting factors for prudential requirements, sustainability benchmarks).

Action 2: Creating standards and labels for green financial products. Building on the future EU sustainability taxonomy, EU standards, and labels for sustainable financial products would protect the integrity of and trust in the sustainable financial market and enable easier access for investors seeking those products. Labeling schemes can benefit retail investors who want to express their investment preferences on sustainable activities.

Action 3: Fostering investment in sustainable projects. Mobilizing private capital for sustainable projects, especially for infrastructure.

Action 4: Incorporating sustainability when providing financial advice. Investment firms and insurance distributors can play a central role in reorienting the financial system toward sustainability by providing advice. Before the advisory process, these intermediaries are required to assess clients' investment objectives and risk tolerance to recommend suitable financial instruments or insurance products. However, investors' and beneficiaries' preferences concerning sustainability are often not sufficiently considered when such advice is given.

Action 5: Developing sustainability benchmarks. Benchmarks are indices that play a central role in the price formation of financial instruments and other relevant assets in the financial system. Benchmarks are helpful instruments for investors, as they allow tracking and measuring of performance and allocating assets accordingly.

Action 6: Better integrating sustainability in ratings and market research. The lack of broadly accepted market standards for corporate sustainability performance assessment renders the transparency of the methodology used by research providers significant. However, the extent to which sustainability factors are being considered remains unclear. The Commission is monitoring developments in the credit rating market. It acknowledges the need for greater understanding and transparency about how credit rating agencies consider sustainability factors. The Commission will invite ESMA to promote solutions, ensuring that credit rating agencies fully integrate sustainability and long-term risks.

Action 7: Clarifying institutional investors and asset managers' duties. The proposal will aim to(i)explicitly require institutional investors and asset managers to integrate sustainability considerations in the investment decision-making process and(ii) increase transparency towards end-investors on how they integrate such sustainability factors into their investment decisions, in particular in terms of their exposure to sustainability risks.

Action 8: Incorporating sustainability into prudential requirements. Banks, insurance companies, and pension funds are the primary sources of external financing for the European economy and an essential channel of savings for investments. Building on the development of the EU sustainability taxonomy, the commission will assess whether more appropriate capital requirements could be adopted to better reflect the risk of sustainable assets held by banks and insurance companies.

Action 9: Strengthening sustainability disclosure and accounting rules. The directive enables companies to report sustainability information flexibly. In the future, an appropriate balance will be needed between flexibility and the standardization of disclosure necessary to generate the data required for investment decisions. In terms of disclosure by the financial sector, there is merit in enhancing the transparency of asset managers and institutional investors, including how they consider sustainability risks and exposure to climate-related risks.

Action 10: Fostering sustainable corporate governance and attenuating short termism in capital markets. Creating more conducive corporate governance to sustainable investment requires (1) developing and disclosing a sustainability strategy, including appropriate due diligence throughout the supply chain and measurable sustainability targets, and (2) clarifying the rules according to which directors are expected to act in the company's long-term interest.

2.4 The Sustainable Finance Framework

Based on the above Action Plan, in 2018, the EU instituted three building blocks for a sustainable finance framework (EUR-Lex, 2021): (a) The first one is the EU taxonomy, manifested as the Taxonomy Regulation, which is designed to establish a robust, science based classification system that offers a standard definition of sustainability for financial and non-financial companies. This classification also serves as a safeguard against greenwashing practices. (b) The second building block, a mandatory disclosure regime for non-financial and financial companies, furnishes investors with information to make well-informed decisions regarding sustainable investments. Disclosure requirements include the environmental and social impact of a company's activities and the business and financial risks it faces due to its sustainability exposures (the 'double materiality' concept). In this context, the EU Commission has adopted a delegated act under the Taxonomy Regulation, which specifies the information to be disclosed by financial and non-financial companies concerning their environmental performance based on the EU Taxonomy. (c) The third foundational element comprises a suite of investment instruments, such as benchmarks, standards, and labels, meticulously crafted to empower financial market participants in aligning their investment strategies with the European Union's climate and environmental objectives. These tools enhance transparency for market participants, facilitating informed decision-making. These instruments are the EU Climate Transition Benchmarks and the EU Paris-aligned Benchmarks, established under the EU Climate Transition Benchmarks Regulation. These benchmarks serve as critical guides for investors to support the transition to a low-carbon economy and achieve alignment with the goals outlined in the Paris Agreement. In Chap. 3 we extend the description of each of these building blocks.

2.5 Advancing the EU's Sustainable Finance Agenda

Since 2021, the European Union has made strides towards a more sustainable future through pivotal developments shaping the landscape of sustainable finance in Europe. These include the EU Green Deal Plan, the 2021 framework that marks a significant evolution in the EU's approach to sustainable finance, and the European Securities and Markets Authority (ESMA)'s instrumental role in propelling the EU's sustainable finance agenda forward.

The European Green Deal

The Commission launched the European Green Deal in December 2019. During its December meeting, the European Council took note, issuing the following statement as part of its conclusions: *The transition to climate neutrality will bring significant opportunities, such as the potential for economic growth, new business models and markets, new jobs, and technological development. (European Council conclusions, December 12, 2019).*

The European Green Deal is a package of policy initiatives aiming to set the EU on a green transition. With the ultimate goal of reaching climate neutrality by 2050, it supports the transition of the EU into an equitable and thriving society with a modern and competitive economy while emphasizing the need for a holistic, cross sectoral approach in which all relevant policy areas contribute to the overarching netzero emissions goal. The package includes policy initiatives addressing climate, the environment, energy, transport, industry, agriculture, and sustainable finance—all strongly interlinked. As part of the Green Deal, the Commission has proposed several sustainable finance initiatives: Investment plans, taxonomy on green investments, and rules on green bonds (European Council, n.d.).

A New Approach for 2021 and Beyond

In a relatively short period, significant strides have been made in laying the groundwork for the EU's sustainable finance framework, and the Commission remains dedicated to finalizing the ambitious goals set out in its 2018 action plan. Since its inception, the landscape has shifted—both in terms of evolving needs and a changing global context—necessitating a fresh chapter in the EU's sustainable finance strategy. In response, the European Commission unveiled its Sustainable Finance Strategy in 2021, identifying four critical domains where further actions are pivotal for empowering the financial system to support the shift towards a sustainable economy robustly:

- Financing the transition to sustainability: This strategy provides the tools and policies to enable economic actors across the economy to finance their transition plans and reach their climate targets and broader environmental targets, regardless of their starting point.
- Inclusiveness: This strategy respects the needs of and provides opportunities for individuals and small to medium-sized companies to have greater access to sustainable finance.
- Financial sector resilience and contribution: This strategy sets out how the financial sector itself can contribute to meeting Green Deal targets while simultaneously becoming more resilient and combatting greenwashing.
- Global ambition: This strategy sets out how to promote international cooperation for an ambitious global sustainable finance agenda.

ESMA's Role in Advancing the EU's Sustainable Finance Agenda

The promotion of the European financial framework discussed above requires the involvement of the European Securities and Markets Authority (ESMA). ESMA is the EU's financial markets regulator and supervisor. Its mission is to enhance investor protection, foster well-regulated financial markets, and ensure the financial system's stability. Specifically, it aims to ensure that the financial markets support and promote the EU priorities, which, in this context, are transitioning towards a greener and more sustainable economy. In February 2022, ESMA developed a roadmap for 2022–2024 (ESMA, 2022), which sets three priorities for sustainable finance:

1. Tackling green-washing and promoting transparency: Investigating the complex issue of green-washing, defining its fundamental features, and taking coordinated action in multiple sectors to find common solutions.
2. Building National Competent Authorities (NCAs) and ESMA's capacities: Training initiatives at national and European levels and sharing supervisory experiences among NCAs.
3. Monitoring, assessing, and analyzing ESG markets and risks: Monitoring, assessment, and supervision to identify risks that may negatively impact investors or the stability of financial markets on a timely basis.

Building on its roadmap for 2022–2024, ESMA has seamlessly transitioned into its strategic vision for 2023–2028, ensuring the continuity of objectives while adapting to the evolving landscape of sustainable finance. This progression reflects a deepening commitment to tackling the complexities of greenwashing, enhancing transparency, and ensuring robust market oversight.

The strategic vision for 2023–2028 (published in September 2022) sets forth a trio of core priorities: enhancing market efficiency and financial stability, reinforcing the oversight of EU financial markets, and amplifying the safeguarding of retail investors. Additionally, the strategy emphasizes two pivotal themes: 'Promoting technological innovation and optimizing data utilization' and 'facilitating sustainable finance.' The sustainable finance component is articulated through four key avenues:

1. Effectiveness and integrity of ESG markets: ESMA will maintain investors' confidence in ESG investments by promoting high-quality sustainability disclosures and reducing the risk of greenwashing.
2. A workable regulatory framework: ESMA will improve the regulatory framework and reduce its complexity, increase data availability, and improve consistency across requirements.
3. Effective and consistent supervision: ESMA will pay particular attention to the implementation and supervision of the regulatory framework.
4. The role of retail investors: ESMA will enhance participation by promoting the development of clear and accessible signals and labels, high-quality investment advice, and easy access to high-quality ESG data to facilitate informed investment choices.

2.6 Concluding Remarks

This chapter describes the major events that led to the European Sustainable Finance Strategy and its changes over the years (see timeline in Fig. 2.1), delineating three levels of institutions: The European Union (EU), the European Commission (EC), and the European Securities and Markets Authority (ESMA). This strategy led to establishing the EU Sustainable Finance Framework and regulations, tools, and standards for the EU member states, which will be described in the next chapter.

The strategic vision established over the years reiterates the importance of building competencies within National Competent Authorities and ESMA itself. It emphasizes the need for a more resilient financial sector that can contribute effectively to sustainability goals. By integrating technological innovation and data utilization, ESMA aims to fortify its approach to sustainable finance, ensuring that its initiatives are both forward looking and grounded in the current needs and challenges of the financial markets.

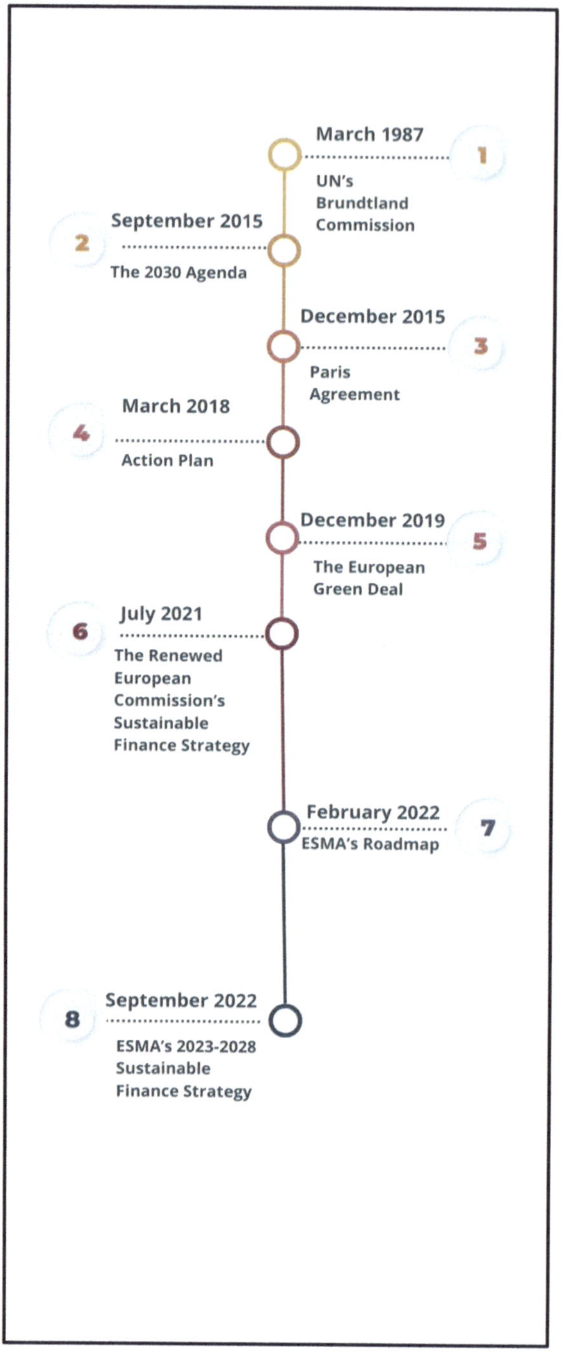

Fig. 2.1 EU Pathways to Sustainable Finance Strategy

References

Bromley, D. W. (2008). *Sustainability: The new Palgrave dictionary of economics* (2nd ed., pp. 293–318). Palgrave Macmillan.

Brundtland, G. H. (1987). What is sustainable development? *Our Common Future, 8*(9).

DiSegni, D. M., Huli, M., & Akron, S. (2015). Corporate social responsibility, environmental leadership and financial performance. *Social Responsibility Journal, 11*(1), 131–148.

ESMA Strategy 2023–2028. (2022, September). https://www.esma.europa.eu/sites/default/files/library/esma_strategy_2023-2028.pdf

EUR-Lex. (2020, June). *Regulation (EU) 2020/852 of the European Parliament and of the Council of 18 June 2020 on the establishment of a framework to facilitate sustainable investment, and amending Regulation (EU) 2019/2088.* https://eur-lex.europa.eu/legal-content/EN/TXT/?uri=celex%3A32020R0852

EUR-Lex. (2021, July). *Communication from the Commission to the European Parliament, the Council, the European Economic and Social Committee, and the Committee of the Regions.* https://eur-lex.europa.eu/legal-content/EN/TXT/?uri=CELEX:52021DC0390

European Commission. (2017, June). *Mid-term review of the capital markets union action plan.* https://ec.europa.eu/info/publications/mid-term-review-capital-markets-union-action-plan_en

European Commission. (n.d.) *Overview.* Retrieved September 14, 2023. https://european-union.europa.eu/institutions-law-budget/institutions-and-bodies/search-all-eu-institutions-and-bodies/european-commission_en

European Council. (n.d.) *European green deal.* https://www.consilium.europa.eu/en/policies/green-deal/

Rodriguez, S. I., Roman, M. S., Sturhahn, S. C., & Terry, E. H. (2002). Sustainability assessment and reporting for the University of Michigan's Ann Arbor Campus, Center for Sustainable Systems, Report No. CSS02-04, Michigan.

United Nation. (n.d.-a) *Transforming our world: the 2030 agenda for sustainable development.* Retrieved September 14, 2023, from https://sustainabledevelopment.un.org/post2015/transformingourworld

United Nation. (n.d.-b) *The Paris agreement.* Retrieved September 14, 2023. http://unfccc.int/paris_

Shirit Katav-Herz is a Senior Lecturer at the Academic College of Tel Aviv-Yaffo, Israel, where she heads the Undergrad Program in Economics and Management. Her research fields are behavioral Economics and Public Economics, focusing on the effect of social norms on consumer's decision-making and the implication on policy.

Rachel Calipha is a Senior Lecturer at the Academic College of Tel Aviv-Yaffo, Israel, where she heads the Capital Market specialty track and the Research Institute of Society and Economics (RISE Impact). Her research fields are strategy and Finance.

Chapter 3
The EU Sustainable Finance Framework

Rachel Calipha and Shirit Katav-Herz

Abbreviations

CSRD	Corporate Sustainability Reporting Directive
DNSH	Doing No Significant Harm
ESG	Environmental, Social, and Governance
EUGBS	European Green Bond Standard
IDD	Insurance Distribution Directive
MiFID II	Markets in Financial Instruments Directive II
NFRD	Non-financial Reporting Directive
PABs	Paris-aligned Benchmarks
PAIs	Principal Adverse Impacts
SFDR	Sustainable Finance Disclosure Regulation
SMEs	Small and Medium-sized Enterprises

3.1 Background

The EU Sustainable Finance Framework represents a comprehensive suite of regulations, tools, and standards meticulously crafted to facilitate the enactment of the EU's ambitious sustainable finance strategy outlined in the preceding chapter. This framework is a cornerstone in Europe's transition towards sustainable development encapsulating pivotal regulations such as the EU Taxonomy Regulation, the Sustainable Finance Disclosure Regulation (SFDR), and the Corporate Sustainability Reporting Directive (CSRD), alongside amendments to pre-existing directives like the Markets in Financial Instruments Directive II (MiFID II) and the Insurance Distribution Directive (IDD). These regulations are designed to engage various

R. Calipha · S. Katav-Herz (✉)
Academic College of Tel Aviv-Yaffo, Tel Aviv-Yaffo, Israel
e-mail: Rachelca@mta.ac.il; Shirit@mta.ac.il

R. Calipha et al. (eds.), *Sustainable Finance Regulation in the European Union*, Sustainable Finance, https://doi.org/10.1007/978-3-031-88392-7_3

stakeholders across the investment spectrum, from entities seeking capital to the investors supplying it and intermediaries facilitating these transactions.

This chapter discusses the above regulations as adjusted to the three building blocks suggested in the 2018 framework (taxonomy regulation, disclosure regulation, and tools and standards). For the disclosure regulation, the chapter also specifies the appropriate stakeholders for each regulation: Issuers, market participants, and financial advisers.

3.2 Taxonomy Regulation

The Taxonomy Regulation was published in the Official Journal of the European Union on June 22, 2020, entering into force on July 12, 2020. It is a classification system that assists companies and investors in identifying "environmentally sustainable" economic activities as sustainable investment decision beneficiaries. Environmentally sustainable economic activities "make a substantial contribution to at least one of the EU's climate and environmental objectives, while at the same time not significantly harming any of these objectives and meeting minimum safeguards." The EU taxonomy facilitates the transition of pollution sectors to technology-neutral and fostering transparency through disclosures for financial market participants and large companies (EC, n.d.-a). The EU Taxonomy serves as a guiding framework rather than a prescriptive checklist for investors; it does not rate the "greenness" of companies nor assess the financial performance of investments. It does not impose mandatory environmental performance standards on companies or financial products. Instead, the Taxonomy functions as a comprehensive tool designed to support investors by clarifying what constitutes environmentally sustainable activities. This allows them to gauge the sustainability level of their investments with greater accuracy. Investors retain the autonomy to make their own investment choices, leveraging the Taxonomy to inform those decisions with an eye towards environmental sustainability.

The EU Taxonomy is expected to catalyze a gradual transition towards sustainability, aligning with the EU's climate and environmental objectives. Furthermore, it is designed to be a dynamic tool, subject to regular updates to reflect evolving standards and practices in environmental sustainability. (EC, n.d -a). Six environmental objectives are at the core of the EU taxonomy:

- Mitigation of climate change
- Adaptation to climate change
- Sustainable use and protection of water and marine resources
- Transition to a circular economy
- Pollution prevention and control
- Protection and restoration of biodiversity and ecosystems

The Taxonomy stipulates four primary criteria for an economic activity to be considered environmentally sustainable. First, it must significantly contribute to at least one environmental objective. Second, it must not cause any significant harm to the other five environmental objectives, a principle known as "doing no significant harm" (DNSH). Third, the activity must adhere to minimum safeguards. Lastly, it must comply with the technical screening criteria outlined in the delegated acts of the Taxonomy Regulation.

The Commission is vested with the authority to adopt delegated and implementing acts, which detail the specific actions that competent authorities and market participants must undertake to fulfill the obligations outlined in the regulation. This mechanism ensures that the regulatory framework can be dynamically adjusted and clarified, facilitating compliance and effectively enforcing the regulation's objectives (EC, n.d.-b). Moreover, to ensure that an economic activity substantially contributes to one of the climate or environmental objectives without causing significant harm to any of the others, the EU sets performance criteria known as *technical screening criteria* in delegated acts. To date, criteria have been established for economic activities that can significantly contribute to climate change mitigation and adaptation (EUR-Lex, 2021). For more details, see the complementary Climate Delegated Act (EC, n.d.-c).

The EU Taxonomy significantly contributes to sustainable finance by providing a unified frame of reference for investors and companies, aiding companies in planning and financing their transition towards sustainability. It plays a pivotal role in safeguarding against greenwashing practices by establishing clear criteria for environmental sustainability. Moreover, it facilitates the acceleration of financing for already sustainable projects and those crucial for the transition to a more sustainable economy. The EU Taxonomy fosters a more transparent, resilient, and forward-looking financial system through these contributions.

The Disclosure Obligations in the EU Taxonomy

Under the EU Taxonomy framework, organizations subject to the Corporate Sustainability Reporting Directive (CSRD) must detail in their annual reports the proportion of their activities that qualify under the EU Taxonomy (referred to as Taxonomy eligibility) and their adherence to the standards outlined in the Taxonomy delegated acts (termed Taxonomy-alignment). While this requirement is mandated explicitly for entities within the CSRD's purview, those outside its scope are encouraged to report this information voluntarily. Such voluntary disclosures can serve multiple purposes, including facilitating access to sustainable finance or supporting other business objectives.

The Disclosures Delegated Act delineates the specifics regarding what information must be disclosed by both financial and non-financial entities about the share of

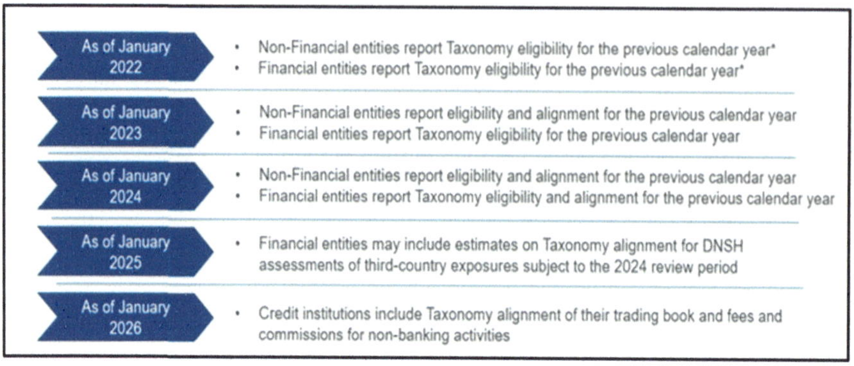

Fig. 3.1 Timeline of reporting by financial and non-financial entities

environmentally sustainable activities within their business operations, investments, or lending practices. It prescribes the content, methodology, and format for these disclosures. Additionally, the Act establishes these entities' reporting requirements and deadlines, as an extension to Article 8 of the Taxonomy Regulation, ensuring a structured approach to reporting on sustainability metrics. The timeline of the reporting by non-financial and financial entities is as follows (Fig. 3.1):

The EU Taxonomy Navigator

To enhance accessibility and practical understanding of the EU taxonomy, the European Commission has developed the EU Taxonomy Navigator. This platform is designed as an intuitive and informative website equipped with online tools to simplify navigation through the EU taxonomy. It features four main tools that assist users in exploring and applying the taxonomy's criteria and guidelines, making the complex framework more approachable and more accessible to utilize for various stakeholders:

- EU taxonomy compass visual representation of sectors, activities, and criteria covered by the EU Taxonomy Delegated Act. https://ec.europa.eu/sustainable-finance-taxonomy/home.
- EU taxonomy calculator-a step-by-step guide on reporting obligations https://ec.europa.eu/sustainable-finance-taxonomy/wizard
- FAQs repository: an overview of questions and answers on the EU taxonomy and its delegated acts. https://ec.europa.eu/sustainable-finance-taxonomy/faq.
- EU taxonomy user guide document guiding non-experts on the Taxonomy https://ec.europa.eu/sustainable-finance-taxonomy/home.

3.3 Disclosure Regulations

The EU disclosure regulations address several stakeholders in the capital market, as described earlier. Below, the regulations are categorized by issuers, market participants, and financial advisers.

Issuers

EU law requires all large corporations and listed companies (except micro-enterprises) to disclose information about perceived risks and opportunities arising from social and environmental issues and about the impact of their activities on humans and the environment. This information can serve investors, civil society organizations, consumers, and other stakeholders in evaluating the companies' sustainability performance in the context of the European Green Deal (EC, n.d.-c). The Non-financial Reporting Directive (NFRD) and Corporate Sustainability Reporting Directive (CSRD) are the two specific corporate laws.

The Non-Financial Reporting Directive (NFRD), enacted in December 2014 and required to be transposed into national law by EU member states by December 6, 2016, marked a significant shift in corporate reporting. Starting from the 2017 fiscal year, with reports first due in 2018, NFRD mandated large companies to disclose information on environmental practices, social and employee matters, human rights adherence, anti-corruption and bribery efforts, and diversity within company boards, including considerations of age, gender, and educational and professional backgrounds. This directive applies to large public-interest entities, encompassing over 500 employees—affecting approximately 11,700 large corporations and groups across the EU, including listed companies, banks, insurance firms, and other entities deemed by national authorities as public-interest entities. This initiative aimed to enhance transparency and accountability, promoting sustainable and responsible business practices across the EU (EC, n.d.-d).

Corporate Sustainability Reporting Directive (CSRD)

In 2022, the European Parliament and the Council of the European Union enacted the Corporate Sustainability Reporting Directive (CSRD), an update and expansion of existing environmental and social impact reporting mandates. This legislation broadens the scope of reporting requirements to include a wider array of large companies and listed small and medium-sized enterprises (SMEs), obligating them to disclose their sustainability practices.

The CSRD aims to provide investors and stakeholders with critical information for evaluating a company's impact on society and the environment, alongside the financial risks and opportunities stemming from climate change and other sustainability concerns. This initiative is expected to significantly enhance transparency, allowing for more informed investment decisions based on sustainability criteria. Anticipating medium—to long-term benefits, the CSRD is projected to lower company reporting expenses by introducing a more standardized reporting framework. This harmonization is crucial for reducing the complexity and costs associated with sustainability reporting. The directive officially came into effect on January 5, 2023. Companies must comply with the CSRD's provisions in their reports starting in 2025, for the financial year 2024, marking a significant step towards integrated corporate transparency and accountability in sustainability matters (EC, n.d.-d).

Market Participants and Financial Advisers

The EU introduced the Sustainable Finance Disclosure Regulation (SFDR) (EC, n.d.-e) as a transparency framework for financial market participants and financial advisers. Financial market participants (EUR-Lex, 2023) are all the financial entities managing their clients' money through financial products. A list of financial products and financial advisers is described in Table 3.1.

Table 3.1 Financial Products and Financial Advisers

(a) Financial Products
Insurance undertaking that provides Insurance-Based Investments Products (IBIPs)
Investment firm that provides portfolio management
Institution for occupational retirement provision
Manufacturer of a pension product
Alternative investment fund manager
Pan-European Personal Pension Product (PEPP) provider
Manager of a qualifying venture capital fund registered under Article 14 of Regulation (EU) No 345/2013
A manager of a qualifying social entrepreneurship fund registered under Article 15 of Regulation (EU) No 346/2013
A management company of an Undertaking for Collective Investment in Transferable Securities (UCITS)
A credit institution that provides portfolio management.
(b) Financial Advisers
An insurance intermediary that provides insurance advice about IBIPs
An insurance undertaking that provides insurance advice about IBIPs
A credit institution that provides investment advice
An investment firm that provides investment advice
An alternative investment fund manager that provides investment advice under Directive 2011/61/EU (on hedge and private equity funds)
An Undertakings for Collective Investment in Transferable Securities (UCITS) management company that provides investment advice under Directive 2009/65/EC on UCITS

SFDR was established in November 2019 and came into application in March 2021. It specifies how financial market participants and advisers must disclose sustainability related information. It helps those seeking investment in entities and projects supporting sustainability objectives make informed choices and properly evaluate how sustainability-related risks are integrated into the decision-making process. SFDR thus contributes to a major EU political objective: attracting private funding to support Europe's transition to a net-zero economy (EC, n.d.-f). In addition, the regulation has double materiality, distinguishing between 'inside-out' and 'outside-in' (Forrest, 2023) and making provisions relevant to both cases.

- Inside-out: Impact of investment decisions on sustainability factors. Firms subject to SFDR must disclose the potentially negative consequences of an investment decision on sustainability factors and the steps taken to mitigate them. Such sustainability factors encompass environmental aspects, e.g., energy performance and water usage, and social aspects, e.g., employee issues and protection of human rights.
- Outside-in: Impact of sustainability factors on investment decisions. Firms must disclose where an ESG event may negatively impact a material investment and adapt their remuneration policies to sustainability risk management.

SFDR further distinguishes between three types of product classification: Products that do not integrate ESG considerations into the investment decision-making process or explain where the integration is not relevant, where products do not meet the criteria of Article 8 or 9 (Article 6); Products that promote environmental and/or social characteristics, amongst other characteristics, and the companies where the investments are made have good governance practices. This implies that ESG investing is not core to these products (Article 8); Products that have sustainable investment as their core objective (Article 9). The regulation also refers to transparency at the entity level and the financial product level.

Financial market participants and advisers are mandated to enhance website transparency by disclosing critical information related to their sustainability practices. This includes detailing how they assess their business models' principal adverse impacts (PAIs), investment decisions, or financial advice on environmental, social, and governance (ESG) factors. Where applicable, they must provide justifications for any determination that their activities do not negatively impact sustainability. Additionally, they must elucidate how sustainability risks are factored into their investment decision making processes and financial advisory services. Moreover, aligning their remuneration policies with integrating sustainability risks is to be communicated, ensuring that their compensation structures support and encourage sustainable investment practices. This disclosure obligation fosters greater accountability and encourages more responsible and sustainable financial practices.

The Sustainable Finance Disclosure Regulation (SFDR) introduces specific transparency requirements for financial products, categorizing them based on whether they promote environmental or social characteristics or are designed to generate a positive environmental or social impact. For both categories, it is mandatory to disclose in pre contractual documents how they intend to achieve their ESG sustainability objectives and to demonstrate in periodic reports the outcomes of these initiatives.

Furthermore, all financial products must outline in the pre-contractual documents how sustainability risks are incorporated into investment decision-making processes and assess the potential impact these risks might have on investment returns. Financial advisers are subject to similar obligations, ensuring that they contribute to the broader sustainability agenda.

Financial market participants who acknowledge the principal adverse impacts (PAIs) of investment decisions on sustainability factors must also clarify whether and how their financial products address these impacts. Financial market participants and financial advisers must clearly articulate in their policies the integration of relevant sustainability risks and make these policies publicly available. This framework ensures a high level of transparency, guiding investors toward more sustainable investment choices and promoting the integration of sustainability considerations into financial advice and product selection.

The Role of MiFID II and IDD Amendments

To ensure that investors' preferences on sustainability and greenwashing concerns are considered, in April 2021, the Markets in Financial Instruments Directive II (MiFID II) and Insurance Distribution Directive (IDD) were amended. The amendments came into effect on August 2, 2022. MiFID II is a legislative framework to strengthen investor protection in the European Union, and the IDD plays a similar role for insurance product customers (Fig. 3.2).

MiFID II and IDD (Nampoothiri, 2023) require advisers to address two investing components—clients and products. For the former (clients), advisers are required to seek their clients' preferences regarding sustainability and, based on them, recommend "suitable" sustainable financial products and insurance-based investment products. For the latter (products), sustainable products should align with the Taxonomy Regulations and SFDR. Clients can determine their sustainability preferences by choosing one or more of the following criteria:

1. A product with a minimum percentage of sustainable investments aligned to the EU Taxonomy (only environmentally sustainable investments currently).
2. A product with a minimum percentage in sustainable investments aligned to the SFDR's Articles 8 and 9 (includes environmental and social assets).
3. The product that considers principal adverse impacts (PAIs).

MiFID II also introduces requirements for product governance. Firms that manufacture or distribute products to clients should identify each product's target market (of potential investors), periodically review it, and ensure that their distribution strategy matches the target market.

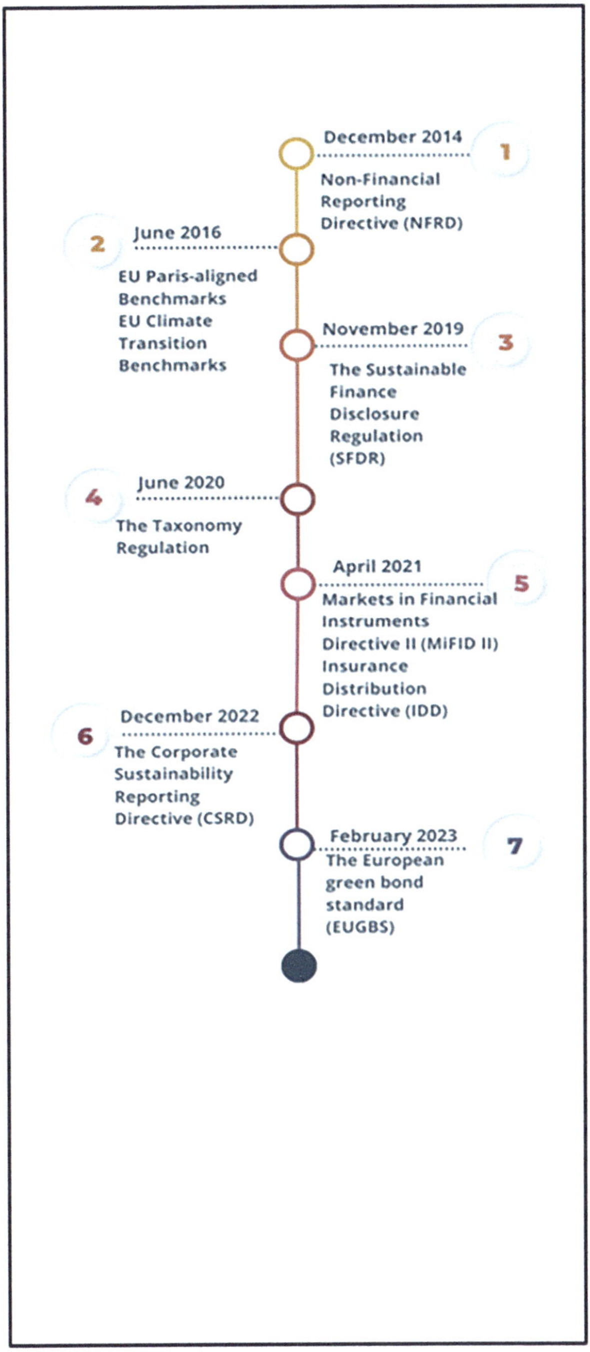

Fig. 3.2 Timeline of the EU Sustainable Finance Framework

Product sustainability factors must be presented transparently. Manufacturers must provide distributors with adequate information about these factors and products, and product distribution should also consider the customer's sustainability objectives. Clients should be informed, where appropriate, of the sustainability factors taken into account when selecting financial instruments in financial advice. In addition, the investment proposal should include a summary of both the advice provided and how the recommendation factors into the client's sustainability preferences. An additional requirement is establishing, implementing, and maintaining risk management policies and procedures incorporating sustainability risk and identifying conflict of interest types.

In summary, the three regulations-EU Taxonomy, CSRD, and SFDR- were created from the ground up, as part of the European Green Deal. Still, the preexistent MiFID II and IDD were amended to conform with the Sustainable Finance Framework (Nampoothiri, 2023). Information is the key to efficient capital allocation to sustainable projects. The regulations define which activities are environmentally sustainable (EU Taxonomy), mandate reporting companies' sustainability performance (CSRD), and provide market participants with guidelines for labeling their financial products based on the EU Taxonomy (SFDR). The two directives, MiFIDII and IDD, instruct financial advisers to integrate this information and act according to their clients' sustainability preferences. The information originates in the corporates and is transmitted to the end investors through market participants and advisers.

3.4 Tools and Standards

The EU created benchmarks (tools) to increase transparency and prevent greenwashing in 2016. In 2020, as part of its Green Deal, it also initiated a discourse of voluntary, high-quality green bond standards.

Benchmarks

The EU established the EU Paris-aligned Benchmarks and the EU Climate Transition Benchmarks in 2016 (Regulation (EU) 2016/1011). Those benchmarks are designed to help investors adopt a low-carbon investment strategy. They aim to improve transparency and comparability, reallocate capital toward climate-friendly investments, and to prevent administrators from making misleading low-carbon claims ('greenwashing') (EC, n.d.-g).

Paris-aligned Benchmarks (PABs) assess investment portfolios' compatibility with the objectives of the Paris Agreement on climate change. The Paris Agreement aims to limit global warming to well below 2 degrees Celsius above pre-industrial levels and to pursue efforts to limit the temperature increase to 1.5 degrees Celsius.

Climate Transition Benchmarks (CTBs) is a novel type of benchmark evaluating the contribution of investment portfolios to the transition to a low-carbon economy. CTBs measure investment portfolio compatibility with the objectives of the EU Climate Law, which aims to achieve climate neutrality in the EU by 2050. Broader in scope than PABs, CTBs reach beyond limiting global warming and focus on shifting to a low-carbon economy. As it is not always transparent to benchmark users whether a particular low-carbon index is aligned with the objectives of the Paris Agreement or merely aims to lower the carbon footprint of a standard investment portfolio, divergent approaches to benchmark methodologies result in the fragmentation of the internal market. To address administrators' potentially illegitimate claims about the low-carbon nature of their benchmarks and to protect investors from confusion and ambiguity regarding the aims and motivation underpinning various categories of so-called low carbon indices used as benchmarks for low-carbon investment portfolios, Member States are liable to adopt their own rules.

Without a streamlined framework that can ensure the accuracy and integrity of the main categories of low-carbon benchmarks applied in individual or collective investment portfolios, Member States' approach may give rise to obstacles to internal markets' smooth functioning. To ensure the effective operation of the internal market for the benefit of investors, enhance its functionality, and guarantee a high degree of consumer and investor protection, it is advisable to amend Regulation (EU) 2016/1011 by introducing a regulatory framework, at the Union level, stipulating minimum criteria for EU Climate Transition Benchmarks and EU Paris-aligned Benchmarks. Ensuring that such benchmarks do not substantially compromise other ESG objectives is crucial. Starting in July 2020, as part of the amended regulation (EU) 2016/1011, CTB and PAB administrators were required to disclose their calculation methodologies, specifying how the underlying assets were selected and weighted, which assets were excluded, and why. To assess the benchmark's contribution to environmental objectives, administrators should disclose how the underlying assets' carbon emissions were measured and report their respective values, including the total carbon footprint of the benchmark and the type and source of the data used.

The European Green Bond Standard (EUGBS)

Green bonds are at the heart of an agreement signed on February 28, 2023, by the European Parliament and EU Council. The European Green Bond Regulation (EC, n.d.-h), an integral part of the European Green Deal, sets a high quality voluntary EU standard for green bonds. The European Green Bond Standard (EUGBS) will be available to companies and entities aiming to secure funds for their green investments, provided they meet rigorous sustainability criteria. In particular, issuers of EUGBS must ensure that a minimum of 85% of bond proceeds are directed towards economic activities in alignment with the Taxonomy Regulation. This will enable investors to readily evaluate, compare, and have confidence in the sustainability of their investments, thereby mitigating the risks associated with greenwashing. The EUGBS is still in process.

3.5 Conclusions

This chapter provides an overview of the European Union (EU) regulatory framework. As illustrated in Fig. 3.2, we observe a continuous evolution in the development and enforcement of various laws. While some regulations have already been implemented, others are still in the process of being developed and adopted. The EU updates the demands regularly and realizes the goal of leading the corporates, the market participants, and the financial advisors towards sustainable decisions. The next chapters present the implementation of this framework, the challenges, and the lessons in five EU members: Spain, Italy, Germany, Estonia, the Netherlands, and Israel.

References

EUR-Lex. (2021, December). *Commission delegated regulations (EU) 2021/2139*. https://eur-lex.europa.eu/legal-content/EN/TXT/?uri=celex%3A32021.R2139

EUR-Lex. (2023, June). *Sustainability-related disclosures in the financial services sector*. https://eur-lex.europa.eu/EN/legal-content/summary/sustainability-related-disclosures-in-the-financial-services-sector.html

European Commission. (n.d.-a). *EU taxonomy navigator*. Retrieved September 18, 2023, from https://ec.europa.eu/sustainable-finance-taxonomy

European Commission. (n.d.-b). *Implementing and delegated acts - Taxonomy regulations*. Retrieved September 18, 2023, from https://finance.ec.europa.eu/regulation-and-supervision/financial-services-legislation/implementing-and-delegated-acts/taxonomy-regulation_en#complementary

European Commission. (n.d.-c). *Complementary climate delegated act*.

European Commission. (n.d.-d). *EU taxonomy for sustainable activities*. Retrieved September 18, 2023, from EU taxonomy for sustainable activities.

European Commission. (n.d.-e). *Corporate sustainability reporting*. Retrieved September 18, 2023.

European Commission. (n.d.-f). *Sustainability-related disclosure in the financial services sector*. Retrieved September 18, 2023, from https://www.finance.ec.europa.eu/sustainable-finance/disclosures/sustainability-related-disclosure-financial-services-sector_en

European Commission. (n.d.-g). *Sustainable finance-minimum standards for climate benchmarks*. Retrieved September 18, 2023, from https://ec.europa.eu/info/law/better-regulation/have-your-say/initiatives/12020-Sustainable-finance-minimum-standards-for-climate-benchmarks_en

European Commission. (n.d.-h). *News*. Retrieved September 18, 2023, from https://ec.europa.eu/commission/presscorner/detail/en/mex_23_130

Forrest, C. (2023). A guide to the EU sustainable finance disclosure regulation for financial institutions. *Anthesis*. Retrieved September 18, 2023, from https://www.anthesisgroup.com/a-guide-to-the-eu-sustainable-finance-disclosure-regulation/#entity

Nampoothiri, M. (2023, April). Introduction of Sustainability Preferences: EU MiFID II & IDD. *NordESG*. https://www.nordesg.de/en/introduction-of-sustainability-preferences-under-eu-mifid-ii-idd

Rachel Calipha is a Senior Lecturer at the Academic College of Tel Aviv-Yaffo, Israel where she heads the Capital Market specialty track and the Research Institute of Society and Economics (RISE Impact). Her research fields are strategy and Finance.

Shirit Katav-Herz is a Senior Lecturer at the Academic College of Tel Aviv-Yaffo, Israel, where she heads the Undergrad Program in Economics and Management. Her research fields are behavioral Economics and Public Economics, focusing on the effect of social norms on consumer's decision-making and the implication on policy.

Part II
National Approaches to Sustainable Finance Implementation

Chapter 4
Spain: Advances and Remaining Challenges in Sustainable Finance

Mikel Larreina and Estibaliz Goicoechea

Abbreviations

AEX 25	Amsterdam Exchange Index 25
AI	Artificial Intelligence
AMCESFI	Autoridad Macroprudencial Consejo de Estabilidad Financiera (Spanish Macroprudential Authority Financial Stability Council)
BME	Bolsas y Mercados Españoles (Spanish Stock Exchanges and Markets)
BSc	Bachelor of Science degree
CAC 40	Cotation Assistée en Continu 40 (Paris Stock Market Index 40)
CIS	Collective Investment Schemes
CNMV	Comisión Nacional del Mercado de Valores (National Securities Market Commission)
COP21	21st Conference of the Parties
COVID-19	Coronavirus Disease 2019
CSRD	Corporate Sustainability Reporting Directive
DAX 30	Deutscher Aktienindex 30 (German Stock Market Index 30)
EBA	European Banking Authority
ECB	European Central Bank
EIOPA	European Insurance and Occupational Pensions Authority
ESG	Environmental, Social and Governance
ESMA	European Securities and Markets Authority
ESRB	European Systemic Risk Board
EU	European Union
FLESB	Forward Looking Exercise on Spanish Banks
FTSE MIB 40	Financial Times Stock Exchange Milano Indice di Borsa 40 (Milan Stock Market Index 40)
GDP	Gross Domestic Product

M. Larreina (✉) · E. Goicoechea
Deusto Business School (DBS), University of Deusto, Bilbao, Spain
e-mail: mikel.larreina@deusto.es; estibaliz.goicoechea@deusto.es

R. Calipha et al. (eds.), *Sustainable Finance Regulation in the European Union*, Sustainable Finance, https://doi.org/10.1007/978-3-031-88392-7_4

IBEX 35	Iberia Index 35
ICAC	Instituto de Contabilidad y Auditoría de Cuentas (Institute of Accounting and Auditing)
ICJCE	Instituto de Censores Jurados de Cuentas de España (Institute of Chartered Accountants of Spain)
ISD	Investment Services Directive
MiFID II	Markets in Financial Instruments Directive II
MiFIR	Markets in Financial Instruments Regulation
MSc	Master of Science degree
NFGS	Network for Greening the Financial System
NFS	Non-Financial Statement
PNFS	Plan Nacional de Finanzas Sostenibles (National Plan of Sustainable Finance)
PRI	Principles for Responsible Investment
PSI 20	Portuguese Stock Index 20
RD	Real Decreto (Royal Decree)
RDL	Real Decreto Ley (Royal Decree-Law)
REA	Registro Economistas Auditores (Registry of Auditing Economists)
RegTech	Regulatory Technology
SupTech	Supervisory Technology
UNESCO	United Nations Educational, Scientific and Cultural Organization
US	United States

4.1 Introduction

According to the scientific consensus, Spain is one of the countries in Europe that will be more affected by the consequences of global warming (IPCC, 2022). Some extreme events like torrential rain and flash floods; *medicanes;* prolonged heat waves and even a rare *derecho* which have occurred in recent years in the Iberian peninsula or its surrounding waters have already been linked to climate change (Alpert et al., 2002; Koseki et al., 2021; Faranda et al., 2022; González-Alemán et al., 2023). This has also been the case for more frequent and dangerous wildfires, the shortening of autumn and winter, the desiccation of wetlands (like UNESCO-protected sites *Doñana* and, even more intensely, *Tablas de Daimiel*), the desertification of the South-East or the loss of glaciers in Andalucia and in the Pyrenees, among others (Miller et al., 2023; Rodrigues et al., 2023).

More frequent extreme events and long heatwaves are directly linked to higher mortality rates (Ermengol et al., 2023), lack of access to water and higher prices (Faus, 2023), the extension of tropical diseases (for instance, the Chikungunya is expected to find favorable conditions in Western Europe), and worsening quality of life, among other societal impacts. Spain (together with Italy and Greece) is also at the forefront of the social consequences of global warming in Europe: current levels of uncontrolled immigration on the southern border of the EU can be expected to increase dramatically when living conditions in the Sahel deteriorate. Extreme right parties all over Europe have capitalized on fear and discontent with growing levels of immigration; there are real risks to democracy if population movements connected to global warming are not appropriately managed (what can be even more challenging if conditions in Northern and Central Africa worsen too fast).

In purely economic terms, the impact of global warming will also be dramatic, as in the rest of the Mediterranean region, in which most socioeconomic sectors are vulnerable to climate change. Besides the impact on coastal and urban infrastructures, it may negatively affect a buoyant industry like tourism. Spain is unfortunately excessively dependent on summer holiday makers looking for sea and sun. Unbearable warmth in the Mediterranean (or persistent droughts) can force a relocation of this kind of tourism towards coasts in more northern locations (Hein et al., 2009; Díaz-Poso et al., 2023). Unless tourism can be deseasonalized or refocused towards wetter and colder regions, the impact will be remarkable, and even in this case, the regions currently focused on tourism would face a severe economic downturn. Mountainous regions depending on ski resorts must reinvent themselves as the snow cover remains for increasingly shorter periods.

Some agricultural industries will also suffer, as a warmer climate will move the geographical boundaries for many species, including relevant agricultural products such as vineyards. Some products must be relocated to new regions (at higher altitudes, for instance), and some areas will become unfit for production. More water stress will negatively affect agricultural production (and result in higher mortality rates in wildlife). Besides, reputational risk will be increasingly present: there have already been boycotts by consumer associations against specific companies or sectors with negative ecological footprints (Wojciechowski, 2023). A rapid transition to a greener economy also has potential advantages: Spain is well placed to produce and export energy using renewable sources and has accordingly developed large industry clusters related to the energy transition in the last decades. More recently, sustainable mobility has become a significant focus of interest, as seen in the projects linked to the Recovery and Resilience Plan agreed between Spain and the EU.

Given this context, it is hardly surprising that citizens in Spain and Portugal show consistently higher climate-change risk perceptions than those of most other European countries (IPCC, 2022). Unfortunately, the polarised socio-political environment of recent years has reduced the pace of legal changes, as the topic is no longer pacific in political terms. Some groups perceive more clearly the negative impacts of climate change; among them are women, younger age groups, more educated, left-leaning, and those with more 'self-transcendent' values (Duijndam & van Beukering, 2021). Regarding ESG Finance, Spain has been active for a long period. There have been private issuers of green and social bonds since 2016, including companies like Iberdrola, the largest renewable utility in the world, which sustainable fund providers fund. When compared to their European peers, Spanish stocks have higher ratings in Environmental issues while performing relatively poorly in Governance areas, as may be seen in the following table (all indices analyzed have significant ground for improvement in Social aspects, though here again, the Spanish Ibex-35 performs better than others). See Table 4.1.

Table 4.1 Average ESG score of companies listed in the selective index Source: Bloomberg and own research. Data from September 2023

Stock Market Index	Stock Exchange	ESG	E	S	G
AEX 25	Amsterdam	56,30	42,05	36,76	89,97
CAC 40	Paris	61,15	49,69	37,49	96,14
DAX 30	Frankfurt	60,77	53,24	41,31	85,49
IBEX 35	Madrid	63,89	59,01	48,37	84,45
FTSE MIB 40	Milan	64,72	55,17	48,26	90,64
PSI 20	Lisbon	51,09	45,32	39,51	68,38
Average		60,65	51,76	42,39	87,39

Finally, the governmental push towards a greener financial system requires the interaction of several bodies, as the supervisory system in Spain is complex. Thus, the financial system is divided into three main areas, each of them with different institutions with regulatory and supervisory powers: Banco de España for banking and related services (as mutual guarantee companies, for instance); CNMV for securities and financial markets; and Dirección General de Seguros y Fondos de Pensiones, the only one which is not independent but a unit within the Ministry for the Economy, for insurance and pension funds. 2008 there was a plan to strip this last supervisor of its powers, although it was not implemented. Furthermore, as the change in the European supervisory framework following the Great Financial Crisis of the late 2000s also established three European Authorities for the domains mentioned above (European Banking Authority (EBA), European Securities and Markets Authority (ESMA), and European Insurance and Occupational Pensions (EIOPA), respectively), it proved helpful in keeping a similar organization in Spain. These European Supervisors have pre-eminence over the Spanish ones: they can directly supervise some Spanish financial institutions (those considered "significant"), have the power to impose regulatory and supervisory requirements, provide mandatory guidelines and interpretations. Another body was created very recently (2019): the Autoridad Macroprudencial Consejo de Estabilidad Financiera AMCESFI (Spanish Macroprudential Authority Financial Stability Council). It replicates the European ESRB at the Spanish level, made after the recommendations of the De Larosière report ten years earlier. It is in charge of addressing systemic (or macroprudential) risk and gathers the top representatives of the Ministry and the three financial supervisors.

4.2 Regulatory Framework

Spain, as a member of the European Union since 1986, is subject to European Union legislation, including in the area of sustainable finance. Hence, Spain has actively participated in the EU's Sustainable Finance Action Plan and has consequently taken steps to promote sustainable finance and address environmental and social issues through legislation and regulatory measures. This journey has not been without turbulence caused by internal politics and external economic shocks, as briefly explained below.

Strategy

Despite the prevalence of European law, Member States have some ground for their legislative developments. For instance, Spain was an early mover in the mid-2000s with an audacious policy of rewarding investments in renewable energies, which still lacked technological efficiency. Unfortunately, the Great Financial Crisis profoundly affected the movements in this direction, shifting the focus of the policymakers, the media, and the public towards recapitalizing the financial system first and then introducing austerity in the public finances, leaving environmental issues for better times (Spain, 2008). This delayed by almost ten years significant moves towards sustainable finance[1] (Larreina & Goitia, 2018). In the case of Spain, the government initially backtracked on its sustainable policies and then turned 180 degrees in 2014, imposing a "*tax on the sun*" which effectively prevented land or property owners from using small devices for self-consumption. A change in government in mid-2018 revitalized environmental legislation and put the environmental emergency at the top of the list of political priorities. Thus, the "tax on the sun" was eliminated that same year, and the right to self-consumption was recognized. Later, the law allowed local energy communities of self-production and self-consumption, in which different households, companies, and administrations can join and share renewable energy, provided their physical location is less than two kilometers away.

Shortly after, Spain formed a multi-ministry workforce to develop an ambitious National Sustainable Finance Plan (Plan Nacional de Finanzas Sostenibles (PNFS)), which operates in parallel to the group tackling global warming led by the Ministry for the Ecological Transition and the Demographic Challenge. The financial task force delivered in 2021 a plan for the Spanish Treasury to issue green bonds, implemented that same year. The PNFS aims to ensure the transformation of both the public and private financial sectors and their adaptation to the risks linked to climate and environmental emergencies while preserving their viability and financial stability. It must also enhance the competitiveness of the Spanish financial sector by

[1] Since the launch of the PRI, Principles for Responsible Investment, in 2006 to the famous discourse of Mark Carney in September 2015 (the tragedy in the horizon), sustainability played almost no role in the significant changes that occurred in the financial system. That same year, Pope Francis I signed the encyclical Laudato Si' (May 2015), where finance is identified as a main enabler of global warming, and the COP21 finished with the signature of the Paris Agreement (December 2015), where again signatories committed to align the financial system to the objective of limiting warming. It is increasingly likely that the most devastating impact of the Great Financial Crisis of 2007–08 in the long term will be that recent efforts to green the financial system did not happen ten years earlier.

leveraging sustainable finance opportunities to create value in the medium and long term. Finally, the plan should help redirect capital flows towards a low-carbon and more sustainable investment economy, promoting the alignment of public investments with the goals of the Paris Agreement.

Despite the clear focus on environmental issues, for the time being, Spain has not gone as far as neighboring Portugal, which, since 2022, has considered "a breach of fiduciary duties" (section 35.3 of Lei de Bases do Clima 98/2021) if financial services firms fail to consider climate risk and climate impact in the short, medium and long term when managing investment portfolios. Though capital markets are increasingly relevant, as in the rest of the EU, Spain remains an economy where companies depend primarily on banking credit for funding. Hence, regulation and supervision of banks are critical to facilitate an orderly transition to a carbon-free economy. The Spanish supervisor, the Banco de España has developed several pilot programs to use AI with Non-Financial Statements (NFS), to categorize them and detect information gaps easily. SupTech and RegTech are likely needed to deal with an ever-growing amount of complex and material information (FSB, 2020). They might be essential for processing large amounts of granular information to monitor how credit is directed toward the investments needed for mitigation and adaptation. Banco de España was not one of the founding members of the Network for Greening the Financial System (NGFS).

However, it joined the network a few months after its launch in 2017 and has since followed NGFS recommendations for addressing climate-related challenges. In 2021, Banco de España initiated top-down supervisory climate stress tests for banks under its macroprudential supervision, utilizing NGFS scenarios for physical and transition risks. Banco de España uses the so-called FLESB (Forward Looking Exercise on Spanish Banks). The main channel for estimating the impact of risks related to global warming is the losses materialized in the credits to non-financial companies, whose business activities can be negatively affected by the transition and physical risks, which, as mentioned earlier, are a significant part of the Spanish GDP. This methodology offers a high sectoral granularity, better capturing the sensitivity to climate-related risks. Banco de España claims that FLESB can project the evolution of credit exposures to the future, which, combined with other risk factors and balance composition, can forecast solvency and profitability ratios for individual institutions and the whole sector. In this sense, the conclusions of the 2022 climate stress test carried out by the European Central Bank among the significant credit institutions of the Eurozone (including 10 Spanish banks) showed that banks do not yet sufficiently incorporate climate risk into their stress-testing frameworks and internal models, despite some progress made in the previous two years (ECB, 2022). Spain and Italy were identified as the largest source of climate-related physical risk for the European banking sector (and consequently, banks from these two countries were identified as most vulnerable to these risks). The model used by the ECB was criticized for seemingly underestimating climate-related risks in Northern countries, as floods or more frequent sea-storms can threaten lowlands in the North more than hilly coastal areas in the South.

Additionally, the Banco de España conducted a comprehensive analysis of climate related risks in collaboration with other financial system supervisors within the country's macroprudential authority (Autoridad Macroprudencial Consejo de Estabilidad Financiera, AMCESFI), comprised of representatives from the Banco de España, the National Securities Market Commission (CNMV), and the Ministry of Economy and Digital Transformation. The report, published in September 2023, shows very general conclusions that reveal the limitations of information that still exists and the complexity in modeling the interaction of economic and climate factors. The institution's commitment to sustainability extends to its investment policies. In February 2021, it joined the Eurosystem agreement to establish a common stance on sustainable investments linked to climate change for non-monetary policy euro-denominated portfolios. Since 2019, Banco de España has incorporated sustainability and responsibility criteria into its investment portfolio. Furthermore, the Banco de España actively invests in green bonds across various currencies and participates in green bond funds. The institution is committed to maintaining and potentially enhancing these sustainability policies. Overall, the Banco de España's multifaceted approach underscores the issue's complexity, though the pace of progress is slow.

The other principal supervisor of the Spanish financial system is CNMV, which is in charge of securities and capital markets. CNMV has picked sustainable finance as one of the four strategic lines for 2023–2024. It aims to revitalize capital markets to promote growth and the transition to a sustainable economy. In this particular context, CNMV has recently surveyed the information provided by companies listed in Spanish capital markets following the mandate of the European Taxonomy (CNMV, 2023). It analyses the information reported by the 98 quoted companies, for which it is mandatory to provide information on the Taxonomy (those with more than 500 employees). The supervisor does not audit the integrity of the information released by the companies, nor does it check if the pace of change in companies' behavior is compatible with the Paris Agreement.

Key Regulations

The current regulation regarding financial markets and instruments in Spain is primarily based on European rules that have been gradually incorporated into Spanish legislation. Some milestones in European legislation are the Investment Services Directive (ISD) of 1993, Markets in Financial Instruments Directive (MiFID) since 2004, MiFID II and Markets in Financial Instruments Regulation (MiFIR), which came into effect in 2018, and the European Taxonomy Regulation adopted in 2020. Simultaneously, some Autonomous Communities[2] like the Basque Country have

[2] In Spain, not all regions or Autonomous Communities have the same competencies. The Basque Country has the most extended level of competencies within Spain.

exclusive competencies in regulating financial institutions and Spanish laws on environmental issues, that has become a matter of controversy between administrations (Arzoz Santisteban, 2009). We start with examining some of the most significant recent legislative developments in Spain's sustainable finance landscape.

Firstly, key laws under review include Law 11/2018, various Decrees on renewable energies, Law 7/2021, along with regulations concerning financial markets and collective investments. We begin with an overview of Law 11/2018, enacted on December 28, which implemented Directive 2014/95/EU and led to substantial amendments in several Spanish legislative frameworks.[3] In addition to other reforms, the legislation introduced significant changes to non-financial reporting requirements in Spain in 2018. From January 1, 2018, onward, Spanish companies have been mandated to include Nonfinancial Statements (NFS) in their annual reporting (Spain, 2018). These statements can be integrated within the management report or presented as a separate document, adhering to the same approval, filing, and publication protocols as the annual accounts. The NFS includes information about the business model, environmental matters, social issues, employee-related matters, respect for human rights, anti-corruption, bribery efforts, and general information about the company. As the threshold to publish the NFS is to have at least 250 employees, 20 listed Spanish companies do not need to report yet on the Taxonomy but do have to release their NFS. Spain is one of the three European Union countries (along with France and Italy) that made it mandatory for the information in the NFS to be verified by an *independent verification service provider* (Goicoechea et al., 2021). Nevertheless, the law does not specify the standard, methodology, or level of verification required, nor does it determine the relevant qualifications the auditor needs. On May 5, 2023, at the request of the Ministry of Economic Affairs and Digital Transformation, the Institute of Accounting and Auditing (ICAC) has initiated a public consultation on the preliminary draft of legislation that will regulate corporate reporting on environmental, social, and governance (ESG) issues in Spain. This draft legislation is designed to incorporate the EU Directive 2022/2464, also known as the 'CSRD Directive,' into Spanish law. The transposition of this directive must be completed by the deadline of July 6, 2024.

Secondly, we discuss the Spanish legislation governing renewable energies, particularly photovoltaic power plants, that has undergone substantial modifications in its objectives over time. Initially, Real Decreto (RD) 436/2004 (Spain, 2004), which established a policy where the government provided premiums for energy produced using solar power, sparking a significant investment boom. By 2008, Spain had become the world's second-largest country in terms of installed solar capacity. However, this rapid growth was curtailed when the financial crisis prompted a reduction in these premiums. Subsequent regulations, including RD 1578/2008, introduced a cap on the maximum amount of premium payable, and RD 1565/2010 further restricted the duration of these payments to 25 years (Spain, 2008; Spain,

[3] Including the Commercial Code, the consolidated text of the Capital Companies Act approved by Royal Legislative Decree 1/2010, dated July 2, and Law 22/2015, dated July 20, on Auditing of Accounts.

2010a). Months later, the reductions also affected pre existing investments (Spain, 2010b). RDL 1/2012 ruled that new investments were no longer eligible for financial incentives that remained strictly limited to pre-existing plants (Spain, 2012), and then, they were canceled for all plants. The change of direction becomes complete when a "sun tax" is approved (Spain, 2015): individuals or entities who generated their electricity were forced to pay a fee to the Spanish government for connecting to the electrical grid. Implementing this punitive "sun tax" was widely criticized for discouraging the use of solar energy and renewable technologies, which are critical for reducing greenhouse gas emissions and combating climate change. In response to the backlash and criticism, the law that effectively punished self-consumption was repealed in 2018 when a new government was formed.

The third key legal development is Law 7/2021 on Climate Change and Energy Transition, which was enacted in May 2021 (Spain, 2021). This law is designed to align Spain with the objectives of the 2015 Paris Agreement. It focuses on facilitating the decarbonization of the Spanish economy and promoting a shift towards a circular economy model that emphasizes the rational use of resources, decent employment and contributes to reducing inequalities.

This Law has two specific goals:

(i) By 2030, to reach a power system with at least 74% of generation from renewable energy sources and to achieve a renewable energy penetration rate in final energy consumption of at least 42%.
(ii) Before 2050, to achieve climate neutrality, as internationally committed, with the power system exclusively based on renewable generation sources.

The Law mandates the design and approval of a National Integrated Energy and Climate Plan (2021–2030) and a National Climate Change Adaptation Plan. Besides, this Law also requires the three sector supervisors of the Spanish Financial System to elaborate together a biennial report on the financial risks related to global warming, on which we will elaborate more later. The information collected by the supervisors follows the mandate of section 32 of Law 7/2021, which establishes the obligation of issuers, credit institutions, insurance companies, and other financial companies to prepare an annual report on the financial impact of risks related to global warming.

The fourth area we want to highlight affects organized financial markets, which are key players in the adaptation and mitigation processes. Stock Exchange and other multilateral trading venues underwent an intense and profound process of transformation and growth since the approval of the Ley del Mercado de Valores in 1988.[4] More recently, a renewed Ley del Mercado de Valores y de los Servicios de Inversión in 2023 accommodates the complexity reached by financial markets, services, and instruments and explicitly recognizes the relations between European Supervisory Authorities and the Spanish regulators, a critical point when most

[4] In the case of Spain, all organized financial markets operate under the umbrella holding BME, which was acquired in 2020 by the Swiss market operator SIX.

guidelines on how information should be reported and monitored will come from the European authorities.

In this same field, a new Real Decreto adapting the regulation for Collective Investment Schemes (CIS) entered into force in November 2023. It states that the management of investment funds and other CISs need to integrate sustainability risks in their procedures, assess them, consider them in the management of the CIS, and ponder the nature, scale, and complexity of the activities. Financial companies must also include sustainability as a source of conflict of interest. It does not go as far as the Portuguese environmental fiduciary duty previously commented, but is an explicit acknowledgment of the issue's relevance.

Finally, there are many other laws and regulations in Spain affecting the financial system.[5] We consider them less relevant than those mentioned above.

4.3 Challenges and Lessons Learned

In our exploration of the Spanish landscape of sustainable finance, we encounter a series of challenges and lessons learned that have shaped the nation's journey toward sustainable financial practices. This chapter discusses the intricacies of these experiences and the strategies employed to promote sustainable finance practices.

Barriers Encountered During the Implementation Phase

The Spanish financial supervisory authorities have been actively evaluating the implementation of legal requirements related to sustainable finance. In recent years, the CNMV conducted reviews of non-financial statements (NFS) published by issuers of securities within the European Union for the years 2020 and 2021. Usually, these reviews were conducted following the requirements in the revised ISAE 3000 review standard and the Guidelines of the Institute of Chartered Accountants of Spain (Instituto de Censores Jurados de Cuentas de España (ICJCE)) or the Registry of Auditing Economists (RAE). Notably, only one issuer had qualifications in its report (none in 2020). The CNMV, in line with ESMA priorities, made numerous recommendations to issuers of NFSs, primarily concerning the following aspects:

(i) Breakdowns of carbon footprint and reduction targets;
(ii) Article 8 Taxonomy Regulation indicators;
(iii) Methodology and concepts used for calculating the gender pay gap, explanation of data and its evolution;

[5] For instance, the Law 7/2022 on Waste and Contaminated Soils, which came into force in April 2022 and includes the promotion of the circular economy as one of its main pillars. It develops a regime of financial guarantees for the correct management of waste (Spain, 2022).

(iv) Scope considerations and exclusions;
(v) Business model and involvement of third parties in the issuer's value chain;
(vi) Consistency between information in the notes to the financial statements regarding provisions and contingencies and that included in human rights and corruption issues in the NFS;
(vii) Variable remuneration linked to sustainability indicators; and
(viii) Impacts of COVID-19.

In 2023, CNMV presented a survey on implementing the Taxonomy by companies listed in organized markets (CNMV, 2023). In 2022, 89 companies were required to include information on the Taxonomy, of which 29% were not eligible for the Taxonomy, as they operate in sectors not included in the European list. Among the report's main findings is that the highest proportion of alignment is reported for investments in fixed assets. In this Key Performance Indicator, there are huge differences within the sample: half of the respondents devote less than 10% of their investments to mitigation or adaptation. At the same time, a quarter of the companies report that more than 75% of their investments are aligned. The latter companies are seemingly investing to make their activities sustainable (or to expand those that already are), while where necessary making investments aimed at transforming their business model and reducing their carbon footprint. CNMV highlights the investing effort of the energy, construction, and real estate sectors to decarbonize their activities. It is noteworthy that despite the impact that global warming will have in Spain, mitigation is the only objective for 75% of the companies, while adaptation is for 7% and both for 14% (with a small percentage not reporting for which objective the investments are aligned).

Banco de España has also surveyed Spanish firms on their view of how climate change and the transition towards a more sustainable economy might affect them: a large majority of smaller firms have still not assessed the impact that climate change and the green transition may have on their activity (Banco de España, 2022). This might preclude them from being able to ask for green loans whose financial conditions will likely be better than those of standard credit.

As mentioned earlier, AMCESFI is a recently created supervisory body interlinked with the three sectorial supervisors. In September 2023, it published the first report on the financial risks related to global warming. Remarkably, other environmental threats are not explicitly mentioned, nor have they been addressed in the report. Despite this being a first exercise, with the logical ground for improvement that it may have, its lecture is sadly disappointing. Some of the analysis is naïve and is not candid enough on the lack of reliable quality data. From the observed survey results (Fig. 4.1), higher values indicate more readiness to deal with climate change, being 100% a situation of complete preparation. The AMCESFI expresses that it looks like issuers of stock are aware of the challenges of climate change and well equipped to deal with them. It briefly comments that 59% of issuers did not respond and that the other 41% has given their self-assessment without any control for its accuracy (AMCESFI, 2023).

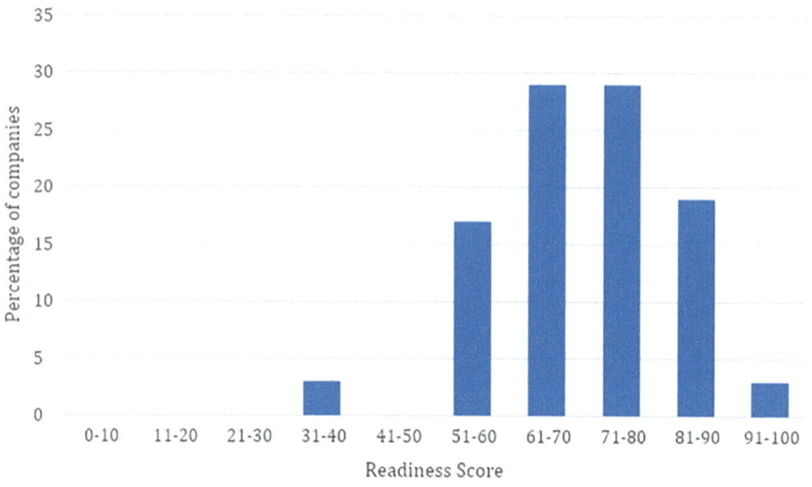

Fig. 4.1 Stock issuers' perceived ability to tackle climate change-related challenges. Source: AMCESFI (2023)

Quite likely, the level of responses shows an excess of confidence, or difficulties to properly understand the question. The questionnaire received the answers of 99 large issuers, with no analysis of the quality of the answers. Besides problems with data quality and relevance, regulators and supervisors face other formidable challenges to properly identify, measure, and manage climate-related risks, and more generally, environmental risks. These challenges include a lack of technical skills in the staff of financial supervisors, a lack of standards in transition plans (their own, and more importantly, those of their clients), and disconnection from risk management frameworks. Regulators and supervisors need to fundamentally upgrade their current toolkits, as they do not have access to the kind of detailed, granular information needed for properly addressing these risks. Besides, the *tragedy of the horizon* (Carney, 2015) consists in the risk extending well over the traditional business and lending cycles traditionally used for assessing risks to financial stability, being thus "invisible" for institutions lacking the ability to interpret their mandates with an environmental perspective.

Addressing the Barriers

The challenges encountered during the implementation of sustainable finance regulations in Spain raise questions about the ability of financial supervisors to analyze complex situations and address their errors. This is exemplified by past experiences, such as the delayed analysis of the 2007–08 financial crisis, published by Banco de España only in 2017. Lessons from other regions acknowledge their failures and shortcomings, contrasting with some of Spain's responses. The National Commission on the Causes Of the Crisis paraphrased Shakespeare, "The fault lies not in the stars, but in us" (National Commission, 2011). Particularly poignant was the Icelandic report, which found that "the Parliament and the Government lacked both the capacity and the courage to set reasonable limits to the financial system" and that the Icelandic financial supervisors did not fully appreciate the serious nature of the matter (Althingis, 2010).

In the aftermath of the crisis, Spanish supervisors did not put much effort into diagnosing the structural deficiencies of the Spanish model (never mind dealing with them). Thus, both before and after the crash, the Spanish central bank, Banco de España, paid much more attention to the workers' wage increases than to the real estate bubble, asset mis-selling or bank mismanagement (Larreina, 2015), while claiming that "there is nothing more imprudent than excessive prudence" (Caruana, 2005). Climate and environmental emergencies require a different approach to regulation.

Strategies Employed to Promote Sustainable Finance Practices

To enhance sustainable finance practices, there must be effective regulation and oversight by regulators and supervisors. As seen earlier, the reviews of the information related to sustainability, including Taxonomy, have checked whether the formal requirements have been met, without analyzing the quality of the data or diving into the details. It is a first step, but more will have to follow soon to make this reporting useful.

In this context, the CNMV has proposed several improvements for the reporting by listed companies. One key suggestion is to provide a clearer explanation of the criteria used to assess eligibility and alignment, including detailed descriptions of the projects, particularly the most relevant ones. Companies might argue that they lack evidence or face significant difficulties in assessing activities for Taxonomy eligibility. In such cases, they may avoid evaluating alignment with the Taxonomy unless the activity is material to the business. This could be a loophole that allows some activities to go unevaluated, necessitating strategies to overcome these evaluation challenges. Additionally, the CNMV recommends increased granularity in the key performance indicators, focusing on the specific economic activities and environmental objectives achieved.

Furthermore, a broader perspective on sustainability is essential. The concept of "sustainable" should not be confined solely to global warming, neglecting other environmental issues, social impacts, and proper governance standards. Sustainability should be integrated across all contexts. For instance, the current political debate includes an extraordinary tax on large banks' profits. Although the analysis by Banco de España remains politically neutral, it does not fully exploit the potential value it could add. Therefore, the supervisor emphasizes the "many areas for improvement" in the design of this tax, yet fails to provide specific suggestions for enhancements, which we will address later in this discussion.

4.4 Future Outlook and Recommendations

In response to evolving environmental challenges, the Spanish Government has been progressively charting a course toward sustainability starting in the year 2000. In the following years, the Spanish Government, having signaled its intention, is expected to present a legislative proposal to Parliament. This proposed regulation aims to mandate, for mid-size and large companies—potentially defined by factors such as employee numbers or financial metrics—the calculation and public disclosure of their carbon footprint across Scope 1, Scope 2, and Scope 3 emissions.

Future Trajectory of Sustainable Finance Regulation in Spain

It is to be noted that since 2014, the government has operated a public register where companies, organizations, and individuals can include information on their carbon emissions, together with their measures for compensating it or their projects for capturing carbon dioxide. For now, some Autonomous Communities like the Balearic Islands have made this disclosure mandatory for large and mid-size companies. Others, like the Basque Country, have approved a legislative project (2024) in which companies increasing their carbon footprint would be taxed at higher rates, and those with high footprints would be penalised when contracting with the administration.

The greater the number of companies mandated to furnish transition plans, the more comprehensive the information available to banks for effectively managing their climate-related risk. Simultaneously, this data empowers regulators and supervisors to discern between financial intermediaries dedicated to a swift and impactful transition towards sustainability and those still supporting businesses reliant on fossil fuels. This distinction is pivotal for promoting responsible and sustainable financial practices.

As mentioned, another forthcoming regulation set to be enacted before mid-2024 is the Law transposing the EU Directive 2022/2464, commonly called the "CSRD Directive." This directive encompasses various aspects, including revising requirements within the corporate reporting framework of Environmental, Social, and Governance (ESG) issues.

Potential Areas for Improvement

An area of improvement is to fully endorse the sensible advice of former Fed chairman Martin and ensure that the supervisor "takes away the punch bowl just when the party was warming up" (Martin, 1955). It is not only a matter of regulation (though robust regulation is needed), but also of having supervisors who believe in their critical role. Additional related recommendations are proposed:

1. Advocate for a shift in the recruitment practices of financial supervisors and institutions. The goal is to establish interdisciplinary teams incorporating scientists, a practice that has begun with some initial steps. Including scientists in these teams will complement the existing skill sets and knowledge of traditional financial intermediaries and supervisors, where finance experts, legal professionals, and computer scientists have historically been the predominant profiles sought after. This transformative step is essential for enhancing the comprehension of climate related risks and their material impact on financial aspects. Furthermore, it will foster a shift in mindset towards forward-looking risk management frameworks, enabling the development of more sophisticated climate stress tests and proactive risk mitigation strategies.
2. Align taxation policies with the imperative of combating and mitigating global warming and advancing broader Environmental, Social, and Governance (ESG) objectives. In addition to refining and enhancing the enforcement of environmental taxes based on the "polluter pays" principle, there is an opportunity to explore innovative approaches that incentivize and support adaptation and mitigation efforts through tax incentives and breaks. For instance, companies that cannot benefit from tax breaks as they do not generate sufficient taxable profits could "transfer" these tax breaks to companies with room for using them, with the commitment that the price paid for the transfer would mandatorily be invested in reducing emissions (or other socially relevant objectives). It may also include different tax rates for financial companies depending on their commitment to fighting global warming and the environmental emergency, as discussed below.

References

Alpert, P., Ben-Gai, T., Baharad, A., Benjamini, Y., Yekutieli, D., Colacino, M., Diodato, L., …, & A. Manes (2002). "The paradoxical increase of Mediterranean extreme daily rainfall despite the decrease in total values". Geophysical Research Letters, 29 (11), 311–314. doi:https://doi.org/10.1029/2001GL013554.

Althingis, R. (2010). *Report of the special investigation commission*. Icelandic Parliament.

AMCESFI. (2023). *Informe Bienal de Riesgos del Cambio Climático para el Sistema Financiero*. Autoridad Macroprudencial Consejo de Estabilidad Financiera.

Arzoz Santisteban, X. (2009). Las competencias autonómicas de supervisión de las entidades de crédito. *Revista de Administración Pública, 178*, 37–87.

Banco de España. (2022). *Informe anual*. Banco de España.

Carney, M. (2015). *The tragedy in the horizon*. Bank of England.

Caruana, J. (2005). *The future of banking supervision in Europe, speech at the conference "supervision of international banks: Is a bank crisis still possible in Europe?"*. European Banking Federation. 28th October. Published in BIS Review 84/2005, BIS.

CNMV. (2023). *Informe sobre los desgloses relativos a la Taxonomía Europea Ejercicio 2022*. CNMV.

Díaz-Poso, A., Royé, D., & Martínez-Ibarra, E. (2023). Turismo y Cambio Climático. Aplicación del Holiday Climate Index (HCI: Urban) en España en los meses de verano para mediados y finales de siglo. *Investigaciones Turísticas, 26*, 274–296. https://doi.org/10.14198/INTURI.23493

Duijndam, S., & van Beukering, P. (2021). Understanding public concern about climate change in Europe, 2008–2017: The influence of economic factors and right-wing populism. *Climate Policy, 21*(3), 353–367. https://doi.org/10.1080/14693062.2020.1831431

ECB. (2022). *2022 climate risk stress test*. https://www.bankingsupervision.europa.eu/ecb/pub/pdf/ssm.climate_stress_test_report.20220708~2e3cc0999f.en.pdf

Ermengol, P. D., Mora, N., Fina, F., Perramon, A., Prats, C., Medina, M., Planella, A., Mompart, A., Mendioroz, J., & Cabezas, C. (2023). Mortality in Catalonia during the summer of 2022 and its relation with high temperatures and COVID-19 cases. *Frontiers in Public Health, 11*. https://doi.org/10.3389/fpubh.2023.1157363

Faranda, D., Bourdin, S., Ginesta, M., Krouma, M., Noyelle, R., Pons, F., Yiou, P., & Messori, G. (2022). A climate-change attribution retrospective of some impactful weather extremes of 2021. *Weather and Climate Dynamics, 3*(4), 1311–1340. https://doi.org/10.5194/wcd-3-1311-2022

Faus, J. (2023). Barcelona raises water bills due to long-running drought. *Reuters*. News published on November 28.

FSB. (2020). *The use of supervisory and regulatory technology by authorities and regulated institutions*. FSB.

Goicoechea, E., Gómez-Bezares, F., & Ugarte, J. V. (2021). Improving audit reports: A consensus between auditors and users. *International Journal of Financial Studies, 9*(2), 25. https://doi.org/10.3390/ijfs9020025

González-Alemán, J. J., Insua-Costa, D., Bazile, E., González-Herrero, S., Marcello Miglietta, M., Groenemeijer, P., & Donat, M. G. (2023). Anthropogenic warming had a crucial role in triggering the historic and destructive Mediterranean Derecho in summer 2022. *Bulletin of the American Meteorology Society, 104*(8), 1526–1532. https://doi.org/10.1175/BAMS-D-23-0119.1

Hein, L., Metzger, M. J., & Moreno, A. (2009). Potential impacts of climate change on tourism; a case study for Spain. *Current Opinion in Environmental Sustainability, 1*(2), 170–178. https://doi.org/10.1016/j.cosust.2009.10.011

IPCC. (2022). *Climate change 2022: Impacts, adaptation, and vulnerability*. Cambridge University Press. https://doi.org/10.1017/9781009325844

Koseki, S., Mooney, P. A., Cabos, W., Gaertner, M. A., de la Vara, A., & González-Alemán, J. J. (2021). Modelling a tropical-like cyclone in the Mediterranean Sea under present and warmer climate. *Natural Hazards and Earth System Sciences, 21*, 53–71. https://doi.org/10.5194/nhess-21-53-2021

Larreina, M. (2015). La Banca Central: de los viejos clichés a los nuevos desafíos (la tragedia en el horizonte). *Boletín de Estudios Económicos, 70*(216), 389–425.

Larreina, M., & Goitia, L. (2018). Nuevos desafíos: la banca y el cambio climático. *Boletín de Estudios Económicos, 73*(224), 251–273.

Martin, W. M. C. (1955). *Address before the investment bankers Association of America on October 19*. Federal Reserve Bank of St. Louis.

Miller, K.A., Santillo, D., & Johnston, P. (2023). *Race against the climate clock: Climate change and extreme weather events in Spain*. Greenpeace Research Laboratories Technical Report (Review) 01. Retrieved from https://es.greenpeace.org/es/wp-content/uploads/sites/3/2023/07/exeter.pdf

National Commission on the Causes of the Financial and Economic Crisis in the United States. (2011). *The financial crisis inquiry report*. U.S. Government Printing Office.
Rodrigues, M., Cunill Camprubí, À., Balaguer-Romano, R., Coco Megía, C. J., Castañares, F., Ruffault, J., Fernandes, P. M., & Resco de Dios, V. (2023). Drivers and implications of the extreme 2022 wildfire season in Southwest Europe. *Science of the Total Environment, 859*, 160320. https://doi.org/10.1016/j.scitotenv.2022.160320
Spain. (2004). Royal Decree 436/2004, of March 12, establishing the methodology for updating and systematizing the legal and economic framework of the activity of electricity production under special regulations. https://www.boe.es/buscar/doc.php?id=BOE-A-2004-5562
Spain. (2008). Royal Decree 1578/2008, of September 26, on remuneration for the activity of electricity production using photovoltaic solar technology for installations after the deadline for maintaining the remuneration of Royal Decree 661/2007, of May 25, for said technology. https://www.boe.es/buscar/act.php?id=BOE-A-2008-15595
Spain. (2010a). Royal Decree 1565/2010, of November 19, regulating and modifying certain aspects related to the activity of electricity production under special regulations. https://www.boe.es/buscar/doc.php?id=BOE-A-2010-17976
Spain. (2010b). Royal Decree-Law 14/2010, of December 23, establishing urgent measures for the correction of the tariff deficit in the electricity sector. https://www.boe.es/buscar/doc.php?id=BOE-A-2010-19757
Spain .(2012). Royal Decree-Law 1/2012, of January 27, which proceeds with the suspension of remuneration pre-allocation procedures and the removal of economic incentives for new installations generating electricity from cogeneration, renewable energy sources, and waste. https://www.boe.es/eli/es/rdl/2012/01/27/1
Spain. (2015). Royal Decree 900/2015, of October 9, regulating the administrative, technical, and economic conditions of electricity supply modalities with self-consumption and self-consumption production. https://www.boe.es/buscar/act.php?id=BOE-A-2015-10927
Spain. (2018). Law 11/2018, of December 28, amending the Commercial Code, the consolidated text of the Companies Act approved by Royal Legislative Decree 1/2010, of July 2, and Law 22/2015, of July 20, on Audit of Accounts, regarding non-financial information and diversity. https://www.boe.es/buscar/doc.php?id=BOE-A-2018-17989
Spain. (2021). Law 7/2021, of May 20, on climate change and energy transition. https://www.boe.es/diario_boe/txt.php?id=BOE-A-2021-8447
Spain. (2022). Law 7/2022, of April 8, on waste and contaminated soils for a circular economy. https://www.boe.es/buscar/act.php?id=BOE-A-2022-5809
Wojciechowski J. (2023). Answer given by Mr Wojciechowski on behalf of the European Commission to the question E-001814/2023 "Boycott by German supermarkets of Huelva strawberries" by Jordi Cañas (Renew), European Parliament. https://www.europarl.europa.eu/doceo/document/E-9-2023-001814-ASW_EN.html

Mikel Larreina is Professor of Financial Institutions and Instruments at Deusto Business School, Deusto University, Spain. Mikel leads the MSc in European and International Business Management (EIBM). His research and teaching are focused on the structure of the financial system, particularly sustainable finance and financial clusters.

Estibaliz Goicoechea is Professor of Accounting and Auditing at Deusto Business School, Deusto University, Spain. Her research and teaching primarily concentrate on corporate information, with a special emphasis on sustainability information and its verification.

Chapter 5
Italy: Sustainable Finance in Italy

Patrizia Riva

Abbreviations

ABI	Associazione Bancaria Italiana
BoSA	Board of Statutory Auditors
BTAR	Banking Book Taxonomy Alignment Ratio
CONSOB	Commissione Nazionale per le Società e la Borsa
CSDD	Corporate Due Diligence Directive
CSRD	Corporate Social Responsibility Directive
EFRAG	European Financial Reporting Advisory Group
ESRS	European Sustainability Reporting Standards
GAR	Green Asset Ratio
ILO	International Labor Organization
ISA	International Standard on Assurance Engagements
KPIs	Key performance indicators
KRIs	Key risk indicators
LSIs	Less Significant Institutions
NFI	Non financial information
NFRD	Non Financial Information Directive
NZBA	Net-Zero Banking Alliance
PRI	Principles for Responsible Investment
RAF	Risk Appetite Framework
TCFD	Task Force Climate-related Financial Disclosures
UNGC	United Nations Global Compact

P. Riva (✉)
Università del Piemonte Orientale, Novara, Italy
e-mail: patrizia.riva@studio-riva.com

R. Calipha et al. (eds.), *Sustainable Finance Regulation in the European Union*, Sustainable Finance, https://doi.org/10.1007/978-3-031-88392-7_5

5.1 Introduction

In Italy, the growing awareness of the effects of climate change and the framework provided by the evolution of the law requirements have increased companies' focus on sustainability. Bank of Italy recently highlighted that between 2021 and the start of 2023, about half of the firms in the industry (excluding construction) and more than one-third of those in services made or planned to make investments with the specific objective of improving energy efficiency and increasing their use or production of renewable energy (Bank of Italy, 2023).

On average, environmentally sustainable investments accounted for 24% of the total investment spending for those firms that made them in the last 2 years. As regards financing, firms have mainly drawn on internally generated resources and loans from banks or other financial intermediaries, and only a minority of firms (17%) relied primarily on public funding, and only 2% used equity or bond venture capital. Environmentally sustainable investments were particularly common among firms (in industries excluding construction) and are positively associated, on the whole, with the intensity with which firms perceive climate risk. In Italy, we find special attention paid to the reliability of non-financial information (NFI) or sustainability reports provided by companies. It is relevant to point out that Italy is one of the three countries, together with France and Spain, which have enacted the obligation of NFI Assurance since 2016 when transposing the NFRD. Furthermore, only Italy structurally requires a higher-level governance system for the processes that lead to drafting such reports, which consists of the supervision provided by the so-called Collegio Sindacale (Board of Statutory Auditors, or BoSA).

5.2 General Italian Regulatory Framework

Firstly, an analysis of the general framework is provided below. To this end, the recent amendments made to the (Italian) Constitutional Law are described below, and an examination of the corporate governance code of listed companies is proposed.

Amendments to the Italian Constitution

Constitutional Law No. 1 of 11 February 2022 amended Articles 9 and 41 of the Italian Constitution, recognizing an express importance to the protection of the environment, both in the part concerning Fundamental Principles and among the provisions of the so-called economic Constitution. In particular, the newly enacted paragraph 3 of Article 9 of the Constitution provides that the (Italian) Republic and, therefore, all Republican institutions: "*(…) shall protect the environment, biodiversity and ecosystems, also in the interest of future generations*". The reference to future generations is essential. The action of public authorities to protect the environment takes on an intergenerational depth, in line with the provisions of other European constitutions, for example, the French, German, Polish, Portuguese, Swedish constitutions, and, even before that, with the principle of sustainable development, which requires pursuing development that ensures meeting the needs of the present generation without compromising the ability of future generations to realize their own needs (Brundtland Report published by the World Commission on Environment and Development in 1987). Article 41 of the Italian Constitution adds: "*Private economic enterprise shall be free. It may not be carried out against the common good or in such a manner that could damage health, the environment, safety, freedom, and human dignity. The law shall provide for appropriate programmes and controls so that public and private-sector economic activity may be oriented and co-ordinated for social and environmental purposes*". These are important and compelling recommendations that outline a precise path of values. The recently amended Article 41 of the Constitution requires considering environmental protection as a prevailing public interest to be imposed on private-sector entities from the outside, shaping their activity and limiting their freedom of economic initiative. Moreover, the wording of the law in legislative terms has imposed a specific objective on companies operating in the domestic context, transforming the environmental and social interest into an objective that must (not may) be pursued by regulated entities, thus putting the theme of sustainability in a new light.

Corporate Governance Code of Italian Listed Companies

In 2020, the Corporate Governance Committee amended the Corporate Governance Code for listed companies, which represents the benchmark for listed companies and is, therefore, a soft law instrument to take into account the issue of sustainability. Right at the opening of the document, more precisely in Principle I of Article 1 "*Role of the Board of Directors*", it is stated that the ultimate goal to be pursued by companies is "sustainable success." This means that a company's governing body's actions must consist of the "*creation of long-term value for the benefit of shareholders, taking into account the interests of other stakeholders that are relevant to the company.*" To understand the scope of this amendment, it is helpful to compare the current wording with the previous one contained in Article 1 of the 2018 Corporate Governance Code. Under Principle 1.P.2, it was stated verbatim that directors should act and make decisions with full knowledge of the facts and autonomously, pursuing "the priority objective of creating value for shareholders in a medium/long-term horizon." No mention was made of the interests of other stakeholders relevant to the company. However, the versatility of corporate objectives that companies that adhere to the Corporate Governance Code must systematically pursue has been made evident. The idea is, therefore, that company management must be inspired by the principles of social responsibility and must take into account both the needs of shareholders and the needs of stakeholders to achieve the ultimate goal of viability by pursuing a complex equilibrium that enables the company to persist in the long term.

5.3 The ESG Disclosure of Italian Companies

A key role, always in the hands of governments, is also played by the availability of high-quality climate-related data and information. Indeed, from a macroeconomic perspective, the availability of such data is a precondition for monitoring financial stability risks and assessing vulnerabilities; from a more micro-economic perspective, ESG reporting increases the companies' awareness and could help them set targets, measure impacts, and govern the change in their organization (Loizzo & Schimperna, 2022; Bank of Italy, 2023).

Non-financial Reporting Requirement for Listed Companies

Legislative Decree No. 254/2016 Implementing Directive (EU) 2014/95 (NFRD)

Directive (EU) 2014/95 (NFRD) represents a milestone towards standardizing corporate reporting of Non-Financial Information (NFI), which is increasingly under the scrutiny of investors and governments to allocate capital and grant businesses the legitimacy and license to operate. On 10 January 2017, Legislative Decree No. 254 of 30 December 2016 was published in the Official Journal, transposing Directive (EU) 2014/95 (NFRD) in Italy. The Decree was entered into force on 25 January 2017, but its provisions became applicable to reports relating to financial years starting on 1 January 2017. Therefore, the first Non-Financial Statements (NFS) were drawn up during the following financial year, i.e., in 2018. According to Article 2 of Legislative Decree No. 254/2016, listed companies are required to respectively draw up an "individual non-financial statement" and a "consolidated non-financial statement" if, during the financial year, the number of their employees was more than five hundred on average and they exceeded at least one of the following thresholds at the reporting date: balance sheet total of €20 million; or, total net revenues from sales and services of €40 million. Individual Non-Financial Statements (as well as the consolidated Statements) must contain information "to the extent necessary to ensure understanding of the business activity, its performance, its results and the impact it produces," at least concerning the following topics: environmental, social and personnel-related; respect for human rights; and fight against active and passive corruption. Article 3 of the Decree also requires the following to be described in the document:

(a) *Business model for managing and organizing the company's activities.*
When describing their business model, companies may consider the opportunity to include appropriate information relating to their business environment, their organization and structure, the markets in which they operate, their objectives and strategies, and the main trends and factors that may influence their future development;

(b) *Policies followed by the company, including due diligence policies and results achieved through them*: any non-financial information disclosed by companies should help investors and other stakeholders understand and monitor the company's performance. The relevant disclosure on policy outcomes may indeed provide useful information on the strengths and vulnerabilities of the enterprise under review;

(c) *Main risks generated or suffered.*
Companies should disclose information on their main risks and how they mitigate them. Such risks may relate to their activities, products or services, their supply chain and business relationships, or other aspects. This may include appropriate scenarios of key short-, medium-, and long-term risks.

Roles of the Auditors & of the (Italian) Board of Statutory Auditors (BoSA)

Article 3, paragraph 7, of the Decree also states that the responsibility for ensuring that a report is drawn up and published in compliance with the provisions of the law lies with the Board of Directors. In fulfilling their obligations, directors must act according to criteria of professionalism and diligence. This requirement is not very different from the one enacted by other European countries. On the contrary, it certainly should be underlined that special emphasis has been given to the issue of reliability of non-financial information in Italy, which immediately imposed a number of assurance obligations on companies required to prepare non-financial reports. Italy, France and Spain are the only countries that have enacted the obligation of NFI Assurance since 2016 when transposing the NFRD. As will be further discussed below, with the future transposition of Directive (EU) 2464/2022 (CSRD), other European countries will also be called upon to be aligned on this point by making assurance compulsory in their turn. It has, in fact, been understood that this is an indispensable requirement in order to obtain greater robustness, thus reliability, of any NFI provided. Furthermore, it should be recalled that only Italy structurally requires a higher-level control system consisting in the supervision provided by the so-called *Collegio Sindacale* (Board of Statutory Auditors, or BoSA). The strict approach of Legislative Decree No. 254/2016 is also demonstrated by the fact that Article 8 imposes a number of administrative penalties and, if the necessary conditions are met, criminal prosecution, which all the above-mentioned parties responsible for drafting and supervising NFI may incur, i.e. directors, members of the Board of Statutory Auditors, as well as Auditors. The assessment and imposition of administrative penalties are activities under the jurisdiction of *Commissione Nazionale per le Società e la Borsa* (CONSOB, National Commission for Companies and the Stock Exchange). Criminal prosecution is under the jurisdiction of Public Prosecutors.

(a) NFI Assurance by the Auditors

As mentioned above, an independent assurance of NFI has been made mandatory in Italy since 2016: the purpose of an Auditor's investigation is both the presence and contents of a Non-Financial Statement. Non-financial statements must be audited to ensure compliance with the relevant requirements. A report must be attached to the Non-Financial Statement and jointly published. The information must be audited and reported to the general shareholders' meeting and the company's annual report. Although for the time being not officially recognized, a standard covering NFI assessments has been set forth IAASB's ISAE 3000 (revised in 2013), Assurance Engagements Other than Audits or Reviews of Historical Financial Information. It is the most widely used principle by professionals engaged in NFI audits (Riva & Bavagnoli, 2022). ISAE 3400 sets out two different levels of assurance: reasonable and limited. These two types of assurance share some key elements. In both cases, assurance providers must clearly define the criteria against which the subject matter will be evaluated, including levels of materiality to be used in assessing the outcome of tests. Providers will adopt the same risk basis for planning their work and delivering a written opinion. However, there are some significant differences. Reasonable assurance is achieved when the risk of a material misstatement of the subject matter has been reduced to a low level. To achieve this, assurance providers must carry out extensive procedures. Assurance providers should obtain sufficient evidence to confirm whether the subject matter meets such criteria. This is essentially the same assurance type as required in an audit of financial statements, where the independent auditor confirms whether the financial statements have been presented fairly. Reasonable assurance conclusions are framed positively. For example, 'based on the procedures performed, in our opinion, the management assertion on [subject matter] has been properly prepared.' To date, for the assurance of sustainability reports/CSR Reports, Europe's most widespread assurance method consists of providing a form of limited assurance. As documented by the European Commission in a survey conducted in 2020 to gather evidence for a review of NFRD, a large proportion of the respondents' non-financial reports that had been assured underwent a limited assurance procedure, i.e., 74% compared to 14% who had undergone reasonable assurance procedures and 11% who had undergone a mixed form of limited and reasonable assurance procedures. Regarding the type of information to be assessed, the most common procedure is to ensure all non-financial reports (i.e., sustainability reports, consolidated non-financial information statements or disclosure, any nonfinancial information published, etc.) and some KPIs, especially regarding Greenhouse gas emissions, energy and waste, employment issues, and the materiality process. In Italy, Commissione Nazionale per le Società e la Borsa

(CONSOB) has chosen to adopt a flexible approach, allowing companies to gradually approach the new requirements and gain the necessary experience over time so that the reporting and internal control systems and external reviews may evolve towards more complex forms at a later time. The current reference level for assessments of Sustainability Reports or NFS is "limited assurance," as such assessments are characterized by the recent enactment of regulatory provisions governing the drafting thereof within a limited scope, by a great quantity of qualitative information, by a scarcity of IT systems for data collection and aggregation, and by internal control systems not yet developed on a par with systems used for the processing of economic and financial data. It is therefore believed that assurance activities, in many cases, cannot be conducted with the necessary level of scope required to be able to express reasonable assurance and positive conclusions, or otherwise, such scope would necessitate a very high effort. In Italy, in particular, there are currently no cases of DNF or sustainability reports assessed with reasonable assurance. However, given the growing importance of sustainability information, companies will be able to request forms of reasonable assurance, at least regarding some particularly relevant types of information, in the form of "mixed" assurance, i.e., with more in-depth analyses, to increase the reliability of such disclosure statements further.

(b) Ex-ante Supervision by the Board of Statutory Auditors (B.o.S.A)

The existence of a Board of Statutory Auditors—BoSA (Collegio Sindacale in Italian) is a peculiar characteristic of the Italian corporate governance system. The Italian Civil Code requires a large number of companies, including SMEs (€4 million in assets or €4 million in revenues or 20 employees), to appoint a Board of Statutory Auditors, BoSA, dedicated to the ex-ante supervision over the board of directors and the company's management actions. In SMEs, this may be the sole auditor. The Board of Statutory Auditors has the task of supervising the Company's compliance with the law and the Bylaws, its conformity to the principles of correct administration, and in particular, the adequacy of its administration, accounting and organizational structure, and its concrete operation. To carry out its activities, the Board of Statutory Auditors, in addition to holding regular meetings and carrying out the controls as mentioned earlier, is required to attend the meetings of the Board of Directors, the company's shareholder meetings, and the executive committee meetings. In a nutshell, corporate control materializes in the following supervisory activities: (1) compliance with the law and the Bylaws; assessment of compliance of the corporate bodies' actions and resolutions with legal provisions and with the provisions of the Bylaws; (2) compliance with the principles of correct administration: assessment of compliance of the directors' operating choices with the general criteria of economic rationality, without delving into the need for, or cost-effectiveness of, such choices; (3) adequacy of the organizational, administrative, and accounting structure adopted by the company and its concrete operation: assessment of the adequacy and operation of company guidelines and procedures, including those relating to the collection and production of financial information. The control tasks over non-financial information delegated to the

Board of Statutory Auditors should be included within the scope of supervisory functions entrusted to the said body according to general principles (Riva & Accettura, 2023). The Board of Statutory Auditors' supervisory work therefore, concerns the adequacy of organizational structures defined by the directors for the pursuit of the strategic objectives identified by the company in social and environmental areas and internal control system and procedures implemented by the directors for risk management purposes and for the collection of data to be included in the NFS. Should such supervisory activities reveal significant risks in regard to the inadequacy of the company's organizational structure, the Board of Statutory Auditors will be required to request the company's CEO, first and foremost, to adopt a corrective action plan in such regard. It will be required to monitor the implementation thereof during its term of office. Associazione per le società per azioni italiane (Assonime, Association for Italian joint-stock corporations) has specified that the Board of Statutory Auditors is required "to perform overall supervisory tasks on systems and processes, which also include non-financial reporting systems and processes, not to verify the correctness of the Non-Financial Statement, but for a different and broader purpose of verifying compliance with the rules of correct administration governing administrative action under the legislation in question."

Expansion of Parties Involved: Implementing CSRD, Directive (EU) 2464/2022

In December 2022, the Parliament and the Council of the European Union adopted the final version of Directive (EU) 2464/2022, known as the Corporate Sustainability Reporting Directive (CSRD), which modified the current reporting requirements of the Non-Financial Reporting Directive (NFRD). The goal of the CSRD is to improve sustainability reporting and contribute to the transition towards a fully sustainable and inclusive economic and financial system in line with the European Green Deal and the United Nations Sustainable Development Goals (SDGs). The new Directive very significantly extends the adoption of sustainability reporting requirements to all large European businesses, banks, and insurance companies, whether listed or unlisted, as well as to all listed companies, with the sole exception of micro-listed companies. The threshold for defining a company as "large" has been set by Accounting Directive 34/2013, i.e., a company should exceed two of the following three criteria at the reporting date to be considered a large company: €20 million in total assets; €40 million in revenues; 250 employees on average per year. In Italy, this means that the number of companies required to prepare a Non-Financial Statement will go from approximately 200 to about 5000. For listed SMEs, the requirement has been postponed until 2026 or, by exercising a specific option, until 2028. For "other SMEs", i.e. unlisted SMEs, there are no formal non-financial reporting obligations. Still, it is worth emphasizing that they may also be affected by the new disclosure requirement as they may be involved in it indirectly. Companies required to submit a report will be called to provide information concerning companies along their production chain to be able to disclose their value chain. The CSRD requires that sustainability information should, as far as possible, cover a company's entire value chain. It is, therefore, reasonable to believe that supply chain leading companies will increasingly be requesting sustainability information from SMEs within their supply chain. Furthermore, an additional indication in this regard is that, as mentioned above, the European Union plans to draft further appropriate standards specifically regarding voluntary sustainability reporting requirements for unlisted SMEs. With the implementation of the CSRD, it will no longer be permitted to draw up a separate report to provide information on sustainability. Indeed, "Recital 57" clarifies that the possibility of publishing a separate report hinders the availability of information that may link financial information to information on sustainability issues. Italian companies will, therefore, be required to disclose information on sustainability in a specific, clearly identifiable section of their report on operations. Among the main innovations introduced by the CSRD, as mentioned above, there is the definition of a European reporting standard that should be commensurate with the level of ambition of the European Union Green Deal and the EU's climate neutrality objective for 2050, in addition to the requirements of other EU legislation and policies, such as the Sustainable Finance Disclosure Regulation and other requirements or recommendations—for example relating to greenhouse gas emissions (EU ETS) or the measurement of environmental performance along the life cycle of products and organizations, as well as the Corporate Sustainability Due Diligence Directive (CSDD), currently being drafted.

The European Sustainability Reporting Standards (ESRS) have been drafted and proposed to the European Commission by the European Financial Reporting Advisory Group (EFRAG). The Framework being developed by the European Union provides for issuing the following three sets of standards, which will therefore become "the" reference standards for Italian companies: general standards, industry standards, and simplified standards for SMEs. As mentioned above, under the CSRD, the assurance of sustainability disclosures will be imposed on all European member states. Italy has already enacted this requirement for some time, as described above. However, it must be emphasized that the CSRD proposal constitutes a path that aims to lead to an assurance level that should be similar for financial reporting and sustainability reporting purposes. European legislators have become aware of the lack of common principles at the European level to certify an entity's compliance with sustainability reporting. Therefore, they have considered it appropriate to adopt a gradual approach to improve the reliability and accuracy of sustainability disclosures, firstly by only requiring limited assurance to be obtained. In particular, assurance must concern compliance with the European Sustainability Reporting Standards (ESRS) in the process of being drafted, compliance with the procedures carried out by companies to identify the information to be disclosed according to sustainability reporting standards (materiality analysis), as well as their compliance with marking obligations. It has thus been established that, by 1 October 2026, the Commission should adopt delegated acts to publish standards aimed at acquiring a limited level of assurance on sustainability reporting compliance, which will establish the procedures that auditors will be required to apply, including the planning of auditing activities, examination of and responses to risks, and description of the conclusions to be included in the sustainability report. The possibility of requiring companies to obtain reasonable assurance will be assessed later.

5.4 Guidelines on Sustainable Finance for the Italian Financial Sector

After considering the legislation on corporate disclosure, this paragraph will focus on the regulatory requirements that govern the Italian financial sector (Arrigoni, 2023, Coppotelli, 2022, De Giudice, 2023). These requirements guide the conduct of financial operations and encourage financial practices referred to as sustainable finance. Table 5.1 summarizes the regulatory and corporate governance documents taken into consideration in this chapter.

The Bank of Italy's Supervisory Expectations on Climate-Related and Environmental Risks

In April 2022, the Bank of Italy published a document entitled *Aspettative di vigilanza sui rischi climatici e ambientali* (Supervisory Expectations on Climate-Related and Environmental Risks), in which it stated that a sustainable growth model should be based on the full integration of environmental, social, and governance (ESG) factors. Renovating the traditional economic growth paradigm, now at the center of the international political agenda, may promote long-term progress that is resilient to external shocks and is, therefore, essential in order to manage the transformations that society and the economic system will have to face in the coming years. In particular, the Bank of Italy has identified the following as relevant: effects of climate change and decarbonization policies; degradation of ecosystems and loss of biodiversity; precariousness and lack of security in the jobs market; and risks linked to low social inclusion and growing inequalities.

Among the ESG components, special attention is paid to environmental issues. This, among other things, results from renewed attention to the so-called United Nations Climate Change Conference Cop26 held in Glasgow in November 2021, which followed the path traced by COP21, Paris 2015, and newly confirmed the objectives set out at that time. At a European level, such objectives materialized in a program dubbed "Green Deal", presented by the European Commission in 2019, under which the Union countries entered into a commitment to reducing greenhouse gas emissions by at least 55. Italy has made these objectives its own by incorporating them into a plan called *Piano Nazionale di Ripresa e Resilienza* (PNRR, National

Table 5.1 Summary of the regulatory and corporate governance documents taken into consideration by the Italian authorities

Summary table of documents examined		
Main sources of Italian legislation	Main sources of EU legislation	State of the art in Italy
1. *Article 41, paragraph 3, of the Constitution (ITA)*: This regulates appropriate plans and controls to guide and coordinate public- and private-sector business activities for social and environmental purposes. 2. *Corporate Governance Code (ITA)*: This provides guidance on the need to pursue sustainable success. 3. *Legislative Decree No. 254/2016 (ITA, transposing the NFRD)*: This introduces the requirement to draft a non-financial statement only for entities of public interest. 4. *Law No. 220 of 9 December 2021 (ITA)*: This introduces a total ban on the financing of companies that, directly or through subsidiaries or affiliated companies, carry out one or more activities related to anti-personnel mines, cluster bombs, and related munitions and sub-munitions, of any nature or composition, or parts thereof. Financial entities are required to define suitable procedural safeguards. 5. *The Bank of Italy's Supervisory Expectations on Climate-Related and Environmental Risks of April 2022 (ITA)*: This defines 12 expectations aimed at integrating climate and environmental risks into the processes and governance of financial entities.	1. *ECB Guide on climate-related and environmental risks of November 2020*: This formulates 13 expectations to integrate emerging climate and environmental risks within the risk management frameworks of financial institutions. 2. *EBA action plan: guidelines on loan origination and monitoring*: This (1) requires the inclusion of ESG factors in credit risk policies and procedures and specific "green" criteria; and (2) requires climate change risks (transitional and physical risks) associated with the borrower to be taken into consideration. 3. *SFDR Directive (Reg. EU 2019/2088)* in force from 2021 defines the concept of sustainable investment and imposes common rules on different categories of financial operators concerning the disclosure of information on sustainability issues. 4. *CSRD Directive (Directive EU 2022/2464)* in force starting from 2024. This extends sustainability reporting requirements to all large companies and sets forth the assurance requirement. 5. *Taxonomy Regulation (Reg. EU 2020/852)* establishes six environmental objectives and the methods through which an economic activity may be classified as sustainable, i.e. whether such activity: (1) contributes to at least one of the goals, (2) without significantly damaging any of the others, (3) meets the technical screening criteria established by the Commission, and (4) is carried out in compliance with the minimum guarantees for the protection of human rights; 6. *Delegated Regulations (EU) 2022/1214 and 2023/363*: Publication of information relating to an entity's exposure to economic activities associated with nuclear energy and fossil gas, as well as the amount and extent of exposure to economic activities eligible for the taxonomy.	1. Findings of the Bank of Italy's November 2022 Survey on climate and environmental risks. Main findings on a sample of less significant institutions (*LSIs* banks) 2. Findings of the Bank of Italy's January 2023 survey on the degree of alignment of Financial Entities with supervisory expectations regarding climate and environmental risks

Recovery and Resilience Plan). The Bank of Italy has emphasized that the ongoing transformation will offer new opportunities and risks for the Italian financial sector.

It is, therefore, essential for operators to prepare suitable safeguards and develop adequate practices to identify, measure, monitor, and mitigate such risks, continuing to guarantee the necessary access to debt and assisting companies engaged in the long and complex transition process with new finance and adequate consultancy services (Anchino et al., 2022; Costa et al., 2022; Di Anonio, 2023; Siani, 2023; Simonetti, 2022). The ability to communicate the integration of climate and environmental risks into one's strategic and operational model is equally important, thereby avoiding incorrect practices (e.g., green-washing), which, on the contrary, would discourage the development of sustainable finance and undermine the operators' reputation. In 2022, the Bank of Italy, as mentioned above, is aware of the fact that climate and environmental risks—which may, in turn, be traced back to traditional financial, credit, market, operational, and liquidity risks—may also have implications for banks and non-banking financial entities under the direct supervision and in line with similar initiatives previously adopted by the ECB and other national supervisory authorities, provided a first set of guidelines entitled "Supervisory expectations." These consider integrating climate-related and environmental risks in corporate strategies, governance and control systems, risk management framework, and disclosure statements by supervised banking and financial entities. Such expectations consist of twelve recommendations of particular importance for operators, as summarized below.

1. *Governance*. The governing bodies of financial entities play an active guidance and governance role in integrating climate-related and environmental risks into the entity's corporate culture and strategy and into its corporate risk appetite framework (where applicable) within the risk limits of the asset portfolios being managed by consistently defining the main company policies and adapting organizational and management systems. In this view, the governing body should approve an appropriate plan of initiatives.
2. *Business Model and Strategy*. When defining and implementing their corporate strategy, financial entities should identify which climate-related and environmental risks can impact their corporate context and understand and measure their impacts to ensure the resilience of their business model and guide their development prospects.
3. *Organizational system and operational processes*. Governing bodies should progressively perform actions involving their organization and operational processes given climate related and environmental risks consistently with and in proportion to assessments made regarding their materiality.
4. *Risk management system*. Financial entities should map any events that may occur because of physical and transitional climate-related and environmental risks and should integrate their risk management system by identifying risks that may be influenced and the prudential implications thereof. *Physical risk* refers to the economic impact arising from the expected increase in natural events whose manifestation may be defined as "extreme" or "chronic". *Acute physical risks* depend on the occurrence of extreme environmental events (such

as floods, heat waves, or droughts) linked to climate change, which may increase the intensity and frequency thereof. *Chronic physical risks*, on the other hand, are determined by climate events that manifest themselves progressively (e.g. gradual rise in temperatures and sea levels, deterioration of ecosystems, or loss of biodiversity). All these types of events may influence the level of production and may even compromise it permanently. *Transitional risk* refers to the economic impact resulting from the adoption of regulations aimed at reducing carbon emissions and encouraging the development of renewable energy, resulting from technological developments and from changing consumer preferences and market confidence.

5. *Availability of an adequate information system regarding climate-related and environmental risks*. The efficiency and effectiveness of an entity's Risk Management system are strictly connected to the availability of an adequate information system from a qualitative and quantitative standpoint. Financial entities should implement actions aimed at creating a comprehensive, high-quality database on climate-related and environmental risk profiles, integrated into an information system suitable for supporting the creation of metrics for the assessment of climate-related and environmental risks.
6. *Integration of risk limitation system*. based on adequate materiality analyses, financial entities should incorporate climate-related and environmental risks into their corporate capital and liquidity assessment processes by supplementing their risk limitation systems.
7. *Need for regular assessment updates*. Given the markedly dynamic nature of climate-related risks, financial entities should outline a plan for regularly reviewing and updating the decisions taken in relation to their assessment methodologies and tools to continuously preserve their validity and significance.
8. *Credit risk*. Financial entities should integrate climate-related and environmental risks into all phases of their credit process, adapting the relevant policies and procedures.
9. *Market risk*. Financial entities should consider the possible impact of climate-related and environmental risks on the pricing of investments in financial instruments, either proprietary assets or assets managed on behalf of third parties, including on a forward-looking basis, in order to minimize the risk of losses. Physical events could lead to changes in expectations, increased risk premiums, increased volatility, and losses in asset value on some markets. Transitional risk factors could generate a sudden repricing of financial instruments linked to sectors affected by environmentally non-sustainable activities. Furthermore, the growing interest of investors in sustainable investment strategies could lead to increased demand for financial instruments considered to be "green" compared to other instruments, thus causing a relative price variation due—if all other factors remain unchanged—to the level of compliance of issuers with the ESG framework.
10. *Operational risks*. Financial entities should consider the possible impact of climate-related and environmental risks on business continuity and on their reputation and legal risk level. Business continuity could be disrupted due to tangible damage to real properties, company branch offices, or data processing centres following extreme climate-related and environmental events.

Furthermore, the evolution of consumer sensitivity regarding climate-related issues may lead to reputation and legal liability risks for financial entities because of the financing of controversial activities from an environmental standpoint. Financial entities may also be exposed to reputation risk if stakeholders perceive that sustainability issues have been promoted exclusively for marketing purposes, without the concrete implementation of actions aimed at supporting the ecological transition.

11. *Liquidity risk.* Financial entities may integrate climate-related and environmental risks into their liquidity risk measurement and management systems by estimating the potential worsening of their liquidity position due to cash outflows and/or decreases in the amount of their reserves and/or changes in the liquidity of internal capital adequacy assessment processes. Financial entities required to draft the Internal Liquidity Adequacy Assessment Process (ILAAP) report must include climate-related and environmental risks in it. Alternatively, such risks will have to be discussed in the Internal Capital Adequacy Assessment Process (ICAAP) report, or in other similar documentation among those included in reporting requirements to which specific financial entities may be subject.
12. *Disclosure to the market.* Financial entities should provide themselves with the infrastructure, data, and processes necessary to disclose the ways in which they integrate environmental risk drivers into their corporate strategy, internal organization, and risk management systems, including the metrics used to assess climate-related risks and sustainability goals.

The above examination reveals that the Bank of Italy's guidelines make reference to the vision previously stated by the European Central Bank (ECB) in its *Guide of the European Central Bank (ECB) on climate-related and environmental risks* (November 2020) regarding the safe and prudent management of climate-related and environmental risks under the current prudential framework, especially when the Bank of Italy describes its expectations regarding how regulated entities should take into account such risks (as drivers for pre-existing risk categories) in the formulation and implementation of their corporate strategies and governance and risk management systems. The Guide also illustrates how, according to the ECB, entities should increase transparency by strengthening disclosures on climate-related and environmental issues. To this end, the ECB has issued a list of thirteen Risk Expectations in order to integrate emerging climate-related and environmental risks into the risk management frameworks of financial institutions. These risks form the basis of the instructions given by the Bank of Italy in the guidelines examined above.

Impact of European Guidelines on Italian Operators

The Bank of Italy's instructions discussed in the preceding paragraph are part of a broader and more complex European regulatory framework which national institutions openly make reference to and operators are consequently required to heed. In particular, in 2018 the European Commission published an "Action Plan for Sustainable Finance", proposing measures aimed at strengthening the role of the financial sector in the creation of a sustainable economy in social and environmental terms. Among the various Action Plan proposals, the entry into force of the Sustainable Finance Disclosure Regulation (SFDR) and of the EU Taxonomy Regulation should be emphasized. The European Commission has also adopted a package of measures aimed at encouraging capital flows into sustainable activities across the Union, including amendments to the delegated acts of the second Markets in Financial Instruments Directive (MiFID II), of the Undertakings for Collective Investment in Transferable Securities (UCITS) Directive, and of the Alternative Investment Fund Managers Directive (AIFMD), integrating sustainability factors and risks into provisions applicable to investment service providers and asset managers, especially in governance and risk management systems. The "EBA Guidelines on Loan Origination and Monitoring" recommending financial entities to incorporate ESG factors and risks associated with them into their credit risk management policies by adopting a fully comprehensive approach came into force in June 2021. In parallel, the EBA published the "Report on ESG Risk Management and Supervision", which sets out a common definition of ESG risks, discusses the transmission channels thereof, and identifies their management and inclusion methods in the regulatory and supervisory framework of banks. On the supervisory level, in November 2020 the ECB published a specific guide on climate-related and environmental risks outlining its expectations in regard of the methods to integrate climate-related and environmental risks into the strategies, business models, governance processes, and risk management frameworks of significant banks, specifying the type of information to be published as part of such public disclosure statement. The main instructions under EU regulations which, as emphasized above, may have an impact on the choices of Italian operators are summarized below.

Adoption of the Sustainable Finance Disclosure Regulation (SFDR)

Italian operators are required to comply with Regulation (EU) 2019/2088, (later supplemented by Regulation 852/2020, in regard of which reference should be made to the following paragraph), regarding Sustainable Finance Disclosure (SFDR). This Regulation constitutes an important reference as it defines the concept of "sustainable investment" (Article 2, 17) and imposes common rules on different categories of financial operators concerning the disclosure of information on sustainability issues.

With reference to *entities,* i.e. financial operators and advisors, the Regulation establishes that they must disclose data on how they take environmental, social, and governance (ESG) factors into consideration at two levels. The processes followed to make investment decisions should be considered first. Subsequently, the assessments made with reference to financial products sold on the European Union markets should also be described. The *Sustainable Finance Disclosures Regulation (SFDR)* requires asset managers to assess and disclose how sustainability risks are considered in their investment processes and how they consider investment decisions that might result in adverse effects on sustainability factors, known as Principal Adverse Impacts (PAIs). More specifically, all financial entities are required to disclose, on their websites, how they integrate sustainability risks into their decision-making processes, how their remuneration policies guiding behaviour are consistent with the integration of sustainability risks, and, finally, how they have identified, examined, and entered a commitment to reducing the adverse impacts of their investment policies on sustainability factors. If a financial entity fails to take into consideration any adverse impacts arising from their choices, they will be required to explain the reason for such circumstances according to the "comply or explain" principle.

With reference to the *scope* of the Regulation, i.e. financial products, the Regulation requires the disclosure of information on how sustainability risks have been taken into consideration in investment decisions and on the consequences they could have on returns. Information should be made available in pre-contractual disclosure statements and the products should also be accompanied by information on the adverse impacts of such investment policies on sustainability factors. It should be emphasized that more detailed information is required regarding products that are defined as "sustainable" products. Two categories of products are identified in the Regulation: (1) "Products that promote environmental and/or social characteristics", described in Article 8, i.e. products which integrate some binding sustainability criteria into the investment process, for example if they apply an exclusions approach or best-in-class strategy; and (2) Products whose objective is sustainable investment, as described in Article 9. Operators who wish to classify their products in one or the other category will be required to clarify how their products meet sustainability characteristics, or how they achieve the objective of sustainable investment. Investors should be able to access data on the operators' website, pre-contractual disclosures, and regular reports.

Taxonomy for the Detection of Sustainable Business Activities

Regulation (EU) 852 of July 2020, as mentioned above, sets six environmental objectives, and describes the methods to classify a sustainable economic activity. This Regulation is also known as *EU Taxonomy Regulation* and outlines the classification framework to determine whether an economic activity is environmentally sustainable, thus requiring the reporting on eligibility and alignment against six environmental objectives.[1] The taxonomy is a common EU-wide classification of economic activities that may be considered environmentally sustainable.

[1] The EU Taxonomy is a classification system that helps companies and investors identify environmentally sustainable economic activities to make sustainable investment decisions. Environmentally

It is conceived as a tool to guide the choices of investors and businesses in view of the transition towards economic growth without adverse impacts on the environment and, in particular, on the climate. Activities are selected according to their ability to contribute to six environmental objectives identified by the European Commission: climate change mitigation; adaptation to climate change; sustainable use and protection of water and marine resources; transition towards a circular economy, including with reference to waste reduction and recycling; pollution prevention and control; and protection of the biodiversity and health of ecosystems (Marucci, 2022). The Taxonomy Regulation also sets out four overarching conditions that an economic activity must meet in order to qualify as environmentally sustainable: making a substantial contribution to at least one environmental objective; doing no significant harm to any of the other five environmental objectives; complying with minimum social safeguards (for example, those set out in the framework agreements of the International Labour Organization—ILO); and, complying with the technical screening criteria set out in the Taxonomy delegated acts.

To facilitate the adoption of the Taxonomy of sustainable investments, the Commission has prepared a Taxonomy Compass which provides a visual representation of the contents of the Taxonomy itself. By using it, operators can evaluate the activities included the objectives to which they contribute, and the criteria they must meet for the purpose of mitigating and managing adaptation to climate change.

Integration of the Taxonomy with Activities Linked to Nuclear Energy and Fossil Gas

The European Commission has assessed that private-sector investment in the gas and nuclear energy sector could play a role in the transition. The activities selected in these two business sectors are in line with the EU's climate-related and environmental objectives and will make it possible to move away from polluting activities more rapidly, such as coal production, in favour of renewable energy sources, which will be the main basis for a future zero climate-related impact. Strict conditions have therefore been established, to which one may add some nuclear and gas activities, as transitional activities, to activities previously included in the first delegated act on climate change mitigation and adaptation. More specifically, strict conditions are required for: gas and nuclear activities which contribute to the transition towards climate neutrality; nuclear activities which meet nuclear energy and environmental

sustainable economic activities are described as activities which "make a substantial contribution to at least one of the EU's climate and environmental objectives", while at the same time not significantly harming any of these objectives and meeting minimum safeguards. The EU Taxonomy is not a mandatory list for investors to invest in. It does not set mandatory requirements on environmental performance for companies or for financial products. Investors are free to choose what to invest in. However, it is expected that over time, the EU Taxonomy will encourage a transition towards sustainability to achieve the EU's climate and environmental goals.

safety requirements; and gas activities which contribute to the transition from coal to renewable energy sources.

Consequently, under Delegated Regulation (EU) 1214/2022 and subsequently under Delegated Regulation (EU) 363/2023 (which amended Delegated Regulation (EU) 2022/1288 of the Commission of 6 April 2022, which had in turn supplemented Regulation (EU) 2019/2088 i.e. the SFDR), operators will be required to publish information relating to: (1) their exposure to economic activities linked to nuclear energy and fossil gas; and (2) the amount and extent of their exposure to economic activities eligible for the taxonomy. In other words, an update was made of the Regulatory Technical Standards (RTS) regarding the content and presentation of information relating to disclosure in pre-contractual documents and regular reports for financial products that invest in sustainable economic activities, with reference to gas and nuclear energy.

Impact of the European Banking Authority

Italian financial operators will be required to consider the instructions of the European Banking Authority (EBA), which published (1) Guidelines in which a road map for the adoption of sustainability principles was described in December 2019; and (2) the Final draft implementing technical standards on prudential disclosures on ESG risks in January 2022.

The former document requires financial companies to implement specific actions in the following direction: (1) inclusion of ESG factors in credit risk policies and procedures by defining "green" criteria; and (2) evaluation of climate-related risks, whether transitional or physical, associated with the financed entity. To effectively implement such instructions, operators are, first, required to consider their own peculiarities in terms of strategy, organizational model, business model, available financial instruments, and business sector to which they belong. Operators should clarify their objectives for the adoption of ESG to apply such goals in operational terms to define a behaviour model, set out a vision, and therefore redefine the entity's corporate culture. Secondly, the EBA refers to Risk Management as a key area for several activities that require a review of processes to integrate ESG factors. These include the definition of a Risk Appetite Framework, i.e. tolerance thresholds, definition of Key Risk Indicators, and monitoring activities over capital requirements.

Finally, financial operators are required to review their credit granting processes. The express intent of the European Commission is to exploit the vehicle of financial operators as a lever to increase the use of sustainable practices throughout the economy. To this end, a recalibration of the banks' capital requirements is also envisaged with the so-called "green support factor", for investments that, being justified in terms of risk and keeping financial stability, are associated with tangible sustainability elements.

In January 2022, the EBA published technical standards on the prudential disclosure of ESG risks, with the aim of expanding the current Pillar 3 framework, i.e. requiring institutions to report qualitative and quantitative information on ESG risks, with a specific focus on climate-related risk, as well as quantitative information on key performance indicators (KPIs) on climate change mitigation measures, including the Green Asset Ratio (GAR) and the Banking Book Taxonomy Alignment Ratio (BTAR). The Green Asset Ratio (GAR) is used to calculate sustainable investments at banks. It represents the share of a bank's taxonomy-aligned economic activities and sustainable investments on its total assets. However, assets that are not aligned with the taxonomy include both assets that are not aligned with the taxonomy and assets whose sustainability cannot be assessed, for example, because they are not covered by the taxonomy or the NFRD/CSRD. As a result, many banks, particularly those focused on providing debt to SMEs, are unable to calculate their GARs in a meaningful way. To provide further information on the extent to which banks finance sustainable activities, the BTAR explicitly includes exposures to entities not covered by the NFRD/CSRD. The challenge for banks is thus to collect the necessary data bilaterally from their customers or to estimate missing data. Nonetheless, a pure focus on GARs would give banks the incentive to lend only to companies covered by the NFRD-CSRD and discourage them from accompanying companies not covered by the NFRD-CSRD in their efforts to transition to carbon neutrality. Italian banks, just as European banks, will have to publish the result of these indicators starting in January 2024 with reference to December 2023.

Ban on the Financing of Production of Anti-Personnel Mines and Cluster Bombs

Finally, it is deemed appropriate to recall here, albeit briefly, the fact that Law No. 220 of 9 December 2021 enacted a total ban in Italy on the financing of companies established in any legal form, with headquarters in Italy or abroad, which, directly or through controlled or associated companies, carry out the construction, production, development, assembly, repair, storage, employment, use, warehousing, stocking, holding, promotion, selling, distribution, import, export, transfer, or shipping of anti-personnel mines, cluster munitions and sub-munitions, of any nature or composition, or parts thereof.

5.5 Implementation in Italy

In this final paragraph, we aim to describe the state of the art in Italy by discussing the findings of two recent surveys conducted by the Bank of Italy.

Findings of November 2022 Bank of Italy Questionnaire Survey

Following the issue, in April 2022, of the document "Supervisory Expectations" examined above, the Bank of Italy conducted a first survey on the extent of integration of climate and environmental risks into the organizational system of the Less Significant Institutions (LSIs) under its direct supervision and distributed to a sample group of 21 Banks (selected to ensure broad diversification in terms of business model, size and geographical area in which the banks operate) a self-assessment questionnaire consisting of four sections: (1) governance and risk appetite, (2) materially assessment, (3) business model and strategy, (4) risk management. As mentioned in the Bank of Italy's report, "Apart from a few positive exceptions, the survey, which covered both transition and physical risks, showed that there is a low degree of alignment with the expectations but, at the same time, it revealed a widespread and growing awareness of the key role of this issue".

In particular, it is reported a diffuse lack of knowledge of ESG matters on the boards of directors, although many banks have established dedicated training programs for their board members; in addition, such knowledge is often also a criterion for evaluating the collective suitability of boards. Another important issue is that a quantitative approach in the measurement of climate risk is still barely used and not fully incorporated in their business model to express ESG targets (in fact, the use of related Key Risk Indicators (KRIs) and Key Performance Indicators (KPIs) is not common) and this is because of the scarcity of data and the lack of IT systems capable of properly handling such data. Most banks (included in the sample) have conducted only a qualitative assessment of the potential impact of environmental factors on the most significant risks to their business models (while a residual proportion of banks conducted no assessment at all). However, the survey also shows increasing attention to ESG issues, and several intermediaries carried out specific analyses aimed at mapping the regulatory framework and/or competitive environment and determining the actions needed to deploy appropriate processes, policies, and strategies(in the short and medium term) to ensure the future sustainability of their business model.

Findings of January 2023 Bank of Italy Questionnaire Survey

Following the first survey focused on the Banks (LSIs), the Bank of Italy also distributed a self-assessment questionnaire to a sample group of 86 non-banking financial entities, including 40 asset management companies, aiming to assess the level of integration of climate-related and environmental risks into their management paradigms, in particular concerning the following areas: (1) governance and organization; (2) business model and strategy, (3) risk management system; and (4) market disclosure. The analysis of responses showed a limited alignment with expectations. Although attention is given to Environmental, Social, and Governance (ESG) issues by senior management generally, in most cases, widespread deficiencies were found, with delays in the implementation, and often also in the planning, of structural actions regarding the various corporate areas involved. Regarding business models and strategies, areas of weakness emerged in assessing how environmental and climate-related risks may influence the competitive and regulatory framework in which each company operates. The majority of financial entities limited themselves to attributing the sustainability of their business models to the presence, in their commercial offer, of products proposed as "green" or "socially responsible." Little attention was paid to the achievement of measurable sustainability objectives as part of their company strategy or to the monitoring of specific performance indicators identified only by a small number of operators. Consequently, the need was emphasized for financial entities who wish to integrate the objectives of combating climate/environmental change into their business models to translate such commitment into measurable sustainability targets.

A differentiated degree of compliance with expectations emerged concerning governance and organizational systems. While for a large part of respondents, the issue of sustainability was being addressed by their company's senior management, there was, however, a non-negligible number of companies in which the degree of involvement of governance bodies was only partial or, in some cases, null. In particular, areas of weakness emerged concerning the lack of expertise on climate-related and environmental issues by governing bodies and an insufficient reporting system. Greater commitment on the part of financial entities in the use of the training lever is desirable: given the complexity of the entities and their continuous evolution—in terms of scientific knowledge and regulatory framework -, training activities must be arranged for and conducted regularly. They should not be limited to sporadic initiatives. Correct monitoring of ESG issues also requires roles and responsibilities within the governing body and/or board committees to be defined and formalized and an effective reporting system to be ensured with adequate frequency.

Finally, the ability of remuneration policies to promote behavior aimed at achieving ESG objectives should be strengthened, including through the identification of specific parameters. About the risk management system, the need for financial entities to speed up their efforts to align with expectations, thus requiring their control units to develop adequate actions, should be emphasized. A widespread problem can be traced back to difficulties in finding reliable and useful data for measuring risks; the quality of information collected mainly from external providers is not being assessed adequately in light of poor data governance strategies and insufficient integration into corporate information systems. Supervisory expectations require the implementation of a systematic integration of sustainability risks into an entity's corporate risk management system.

To this end, it will be necessary for financial entities to complete their mapping of risk events that could occur concerning climate-related and environmental factors and evaluate the materiality and prudential implications thereof. Once a full mapping of these risks has been completed, the acceptable exposure must be determined by defining an adequate system of risk thresholds and indicators, supplementing the

Risk Appetite Framework, if appropriate. In summary, the respondents' general awareness of the need for adaptation has not been translated into sufficiently outlined plans.

Required Action-Plan to All Financial Entities

In light of the findings of the investigations conducted, the Bank of Italy required all financial entities to prepare a suitable "Action Plan" to incorporate consistently the aforementioned expectations of the Authority. In particular, the Bank of Italy has required the governing bodies of LSI banks "to approve a suitable action plan for the next three years to set out a gradual alignment process with the expectations." These action plans will benchmark the supervisory dialogue with the LSIs. More specifically, the Bank will continue to examine and monitor the progress of the initiatives adopted and planned by the intermediaries involved in the survey, which are expected to approve the plan and provide it to the Bank by 31 January 2023 unless they have already done so. At the same time, discussions will begin with the banks not involved in the survey, whose deadline for producing an action plan is 31 March 2023.

The Bank of Italy has also required an Action Plan for the other non-banking financial entities to include: (a) self-assessment of the degree of alignment of company practices with the Expectations; (b) assessment of the materiality of exposure to climate-related and environmental risks based on the specific business model adopted and size and complexity of the company; (c) identification of specific actions to be implemented to fill the gaps found; and (d) definition of priorities and time necessary to complete the various initiatives, taking into account the intensity of exposure to risks and depending on the size and complexity of company operations. After drawing up an "Action Plan" approved by the Boards of Directors of each financial entity involved, accompanied by a report from the Board of Statutory Auditors (BoSA), the same was submitted to the Bank of Italy (by the end of the first quarter of 2023). The Supervisory Review and Evaluation Process (SREP) that the Authority will conduct regarding all supervised financial entities will be taken into consideration on an annual basis.

Good-Practices as Benchmarks for Financial Entities

To support financial entities in the preparation of their "Action Plan," during its investigation, the Supervisory Authority identified some "good corporate practices" for the integration of ESG risks. Despite the weaknesses illustrated above, the analysis of the questionnaires, for example, made it possible to find out that: (a) concerning business models and associated strategic risk, some financial entities had carried out an analysis of their business context and an assessment of the materiality of climate related risks, as a key prerequisite for proceeding with the definition of sustainability strategies. In this regard, the good practice of implementing a monitoring system on the degree of achievement of ESG objectives defined during strategic planning through the identification of specific KPIs also emerged; (b) concerning governance and control systems, some financial entities had prepared a particular sustainability policy by defining the relevant responsibilities in such regard, directly attributing duties to the Board of Directors or creating special organizational structures. Other companies had provided training programs for employees and managers and had included measurable, not merely formal, sustainability objectives in their remuneration policies; and (c) in the area of risk management, some financial entities had started, or at least planned, an initial mapping of the effects of climate-related and environmental risks on their business to evaluate the impact thereof on the various corporate areas. The findings of these analyses will form the basis for introducing sustainability risks into the Risk Appetite Framework (RAF) and capital adequacy self-assessment processes.

5.6 Case Study: Mediobanca Group Approach to Sustainability

We follow with a unique case study of Mediobanca Group. The Mediobanca Group is the fifth Italian bank that is considering capitalization and a premier specialized financial group offering Wealth Management, Consumer Banking, and Corporate Investment Banking services. Its impeccable reputation in Italy results from nearly eight decades of success driven by a responsible approach to business and services that meet the highest standards of excellence. The Group has adopted a Code of Ethics and a Code of Conduct, which define the basic principles to help defend its reputation and set out the values on which its daily operations are based. The Policy is divided into four sections based on areas considered to be priorities: measures to tackle bribery and corruption, human rights, diversity and inclusion, and climate change and the environment. The sustainability approach of Mediobanca Group is detailed in Box 5.1.

Box 5.1 Growth and Sustainability Are Features of Mediobanca's DNA

The Group's strategy is based on the conviction that ethics and profit are not necessarily in opposition because long-term economic growth has to go hand-in-hand with social and environmental progress. To promote a culture geared towards business ethics, integrity, and sustainability, the Group has adopted a Code of Ethics and a Code of Conduct, which define the basic principles to help defend its reputation and set out the values on which its daily operations are based. Mediobanca has also adopted a Group Sustainability Policy to consolidate this approach to govern its direct impacts and identify roles and responsibilities. The Policy is divided into four sections based on areas considered to be priorities: measures to tackle bribery and corruption, human rights, diversity and inclusion, and climate change and the environment. It is based on the principal reference declarations and regulations, including the Universal Declaration of Rights, the ILO's Declaration on Fundamental Principles and Rights at Work, and the Rio Declaration on the Environment and Development. It applies to the whole Group in the countries where Mediobanca operates, and the Group also encourages its customers and suppliers to adopt the Policies principles and provisions. In line with the roadmap, which sets out the Group's objectives in the area of diversity and inclusion, the Mediobanca Group Diversity, Equity, and Inclusion Code has been adopted to promote an inclusive corporate culture in which diversity and equity are defended, without any form of discrimination, including in external relations. With a specific focus on Environment and Climate Change, which is another of the priorities stated in the Group Sustainability Policy, awareness of the fundamental role of finance in the transition to a net-zero emissions economy has led Mediobanca to launch a series of initiatives aimed at raising awareness throughout the organization at all levels—from the Board of Directors and its Committee to the entire personnel—on risk and opportunities linked to and arising from climate change. Climate change and the other topics on which the Group's indirect social and environmental impacts are relevant are governed in the Group ESG Policy, which defines the reference principles and the positive and negative screening criteria applicable to lending, proprietary investing, and investment advisory services to clients. In defining the Policy, reference was made to internationally recognized principles, such as the Universal Declaration of Human Rights, the ten principles of the UN Global Compact, and the United Nations Sustainable Development Goals (SDGs).

To ensure sustainability is more firmly integrated into the company's operations, the Group set a series of ESG objectives (all of which have been comfortably delivered), qualitative and quantitative, in the 2019–2023 Strategic Plan. As further confirmation of the desire to ensure that sustainability is firmly integrated into the Group's industrial strategy, the new 2023–2026 Strategic Plan contains new quantitative ESG targets, also included in the staff performance evaluation and remuneration policies for the corporate population, and identifies a series of cross-divisional actions to help contribute towards a more sustainable future. In the 2023–2026 Strategic Plan, Mediobanca reiterated its climate commitment to achieve carbon neutrality by 2050, with the interim target of a 35% reduction in financed emissions intensity by 2030. At the same time, the Group maintained its focus on direct impact, confirming the use of electricity 100% from renewable sources by

2026 and maintaining carbon neutrality on its emissions. Coherently with its ambition of aligning its business activities with the goals of the Paris Agreement, Mediobanca formalized its support to Climate-related Financial Disclosures (TCFD) and joined the Net-Zero Banking Alliance (NZBA), committing to reduce greenhouse gas emissions arising from its credit and investment portfolio at the path required to achieve a net-zero economy by 2050. There are only two international associations and initiatives the Group participates in to consolidate further its contribution to growth, which is responsible and sustainable:

- *United Nations Global Compact.* United Nations initiative to get businesses and firms worldwide to adopt sustainable and socially responsible policies and to report on their implementation. The Group is a member of the Global Compact and a participant in the Global Compact Italian Network.
- *Principles for Responsible Banking (PRB).* The Mediobanca Group signed up to the PRB in April 2021. The PRB was launched by the United Nations in 2019 and has the aim of incentivizing banks to set objectives for sustainable growth aligned with those of the UN 2030 Agenda and the Paris Agreement.
- *Principles for Responsible Investment (PRI).* Mediobanca SGR, RAM AI, and Polus Capital Management have all signed up to the PRI, a United Nations Supported international network to promote sustainable and responsible investment by institutional investors by integrating sustainability considerations into investment decisions and shareholder activism.
- *Net-Zero Banking Alliance (NZBA).* The NZBA is an alliance set up by the United Nations Environment Programme Finance Initiative (UNEP FI), to align loan and investment portfolios with the target of achieving net zero emissions by 2050 in line with the Paris Agreement.
- *Net Zero Asset Managers Initiative (NZAMI).* Initiative, to which the Group is a signatory via RAM, which brings together an international group of asset managers committed to supporting the goal of net zero greenhouse gas emissions by 2050 or sooner, in line with global efforts to limit warming to 1.5 °C.
- *Task Force On Climate Financial Disclosure (TCFD).* In April 2022 Mediobanca adhered to the recommendations made by the Task Force on Climate-related Financial Disclosure set up by the Financial Stability Board (FSB), to disclose and transparently represent information on its environmental impact through the TCFD Report.
- *Valore D.* Mediobanca participates in Valore D, the first Italian association of large businesses set up to support diversity and women in leadership.

5.7 Summary Table of Documents Examined

The following table summarizes the regulatory and corporate governance documents taken into consideration in this chapter.

References

Anchino, S., Ciavarella, A., Deriu, P., Fiore, F., Nocella, S., Tambucci, M., & Turi, A. (2022). *Gestione del risparmio e sostenibilità: l'approccio dei gestori in Italia*. CONSOB.
Arrigoni, M. (2023). Finanza sostenibile, servizi di investimento e tutela dell'investitore. *Rivista di Diritto Bancario*.
Bank of Italy. (2023). *The uptake of environmentally sustainable investments in 2021–22*. Survey of Industrial and Service Firms.
Bank of Italy. (2023). *Rischi climatici e ambientali. Principali evidenze di un'indagine tematica condotta dalla Banca d'Italia su un campione di intermediari finanziari non bancari*.
Coppotelli, C. (2022). *La strategia europea sullo sviluppo sostenibile. In particolare, la finanza sostenibile e le modifiche al quadro regolamentare europeo*. il Mulino.
Costa, D., Gentile, M., & Linciano, N. (2022). *Interesse verso gli investimenti sostenibili. Un esercizio di caratterizzazione degli investitori italiani sulla base delle indagini CONSOB*. CONSOB.
De Giudice, A. (2023). *La finanza sostenibile. Strategie, mercato e investitori istituzionali*. Giappichelli.
Di Anonio, M. (2023). *ESG transition in banks: New strategies and the impact on management and control systems*. ABI.
Loizzo, T., & Schimperna, F. (2022). *ESG disclosure: Regulatory framework and challenges for Italian Banks*. Questioni di Economia e Finanza n. 744/2022, Bank of Italy.
Marucci, M. (2022). *Finanza sostenibile e tassonomia europea. Una spinta verso l'economia verde*. INAPP.
Riva, P., & Accettura, S. (2023). La vigilanza del collegio sindacale sulla rendicontazione di sostenibilità. AIDEA XL National Conference.
Riva, P., & Bavagnoli, F. (2022). *Limited or reasonable assurance for NFI? Effectiveness and criticalities*. Department of Economics and Business Studies, Università del Piemonte Orientale.
Siani, G. (2023). *Il risparmio gestito: prime evidenze sulla gestione dei rischi Marucci Mclimatici e ambientali*. Bank of Italy.
Simonetti, L. (2022). *Finanza sostenibile e banche centrali. Nuovi attori per una geografia finanziaria in evoluzione*. Bollettino della Società Geografica Italiana.
World Commission on Environment and Development. (1987). *Our common future. Brundtland Report*. United Nations.

Patrizia Riva is an Associate Professor at University' del Piemonte Orientale, Italy, and teaches courses in Corporate Governance, Auditing, and Strategy.

Chapter 6
Sustainable Finance in Germany: Trends and Opportunities

Saadat Hussain

Abbreviations

BaFin	Bundesanstalt für Finanzdienstleistungsaufsicht (Federal Financial Supervisory Authority)
CSRD	Corporate Sustainability Reporting Directive
ECB	European Central Bank
FNG	Forum Nachhaltige Geldanlagen
GSIA	Global Sustainable Investment Alliance
GTAI	Germany Trade and Invest
ISSB	International Sustainability Standard Board
KfW	Kreditanstalt für Wiederaufbau (Credit Institution for Reconstruction)
NGFS	Network of Central Banks and Supervisors for Greening the Financial System
SFB	Sustainable Finance Beirat (Sustainable Finance Advisory Committee)

6.1 Introduction

Germany's financial strength is globally acknowledged, solidifying its position as the largest economy in the European Union and the fourth largest worldwide. The country constitutes 25% of the EU GDP (Germany Trade and Invest, GTAI, 2022a; International Trade Administration, 2023) and hosts 84.6 million people (Federal Statistical Office, 2024), making it the largest market in Europe and a crucial gateway to the EU. The financial industry in Germany is recognized as an attractive market with a robust financial infrastructure and an AAA rating. Financial and

S. Hussain (✉)
Ruprecht-Karls-Universität Heidelberg, Heidelberg, Germany
e-mail: saadat.hussain@csi.uni-heidelberg.de

R. Calipha et al. (eds.), *Sustainable Finance Regulation in the European Union*, Sustainable Finance, https://doi.org/10.1007/978-3-031-88392-7_6

insurance activities contribute nearly EUR 120 billion to Germany's GDP, representing 70% of the country's service sector (GTAI, 2022a).

Diversity characterizes Germany's banking sector, extending beyond private banks to include public and cooperative banks. Key players in this sector include Deutsche Bank, Commerzbank, DZBank, and local savings banks, supported by the renowned universal banking system. Additionally, KfW Bank, a government-owned development bank, is crucial in financing economic development, infrastructure, and sustainability projects. Germany dominates the reinsurance sector in Europe and ranks second globally. The insurance industry, covering life, health, reinsurance, accident, and indemnity, is robust, featuring major players like Allianz and Munich Re. Key players shaping Germany's financial landscape also include the Frankfurt Stock Exchange. As a cornerstone of Germany's financial market, it facilitates trading stocks and other financial instruments. Frankfurt plays a unique leadership role with a network of over 200 credit institutions, 7900 financial service companies, and significant roles in Euro clearing and derivatives trading (GTAI, 2022a). The future of the financial sector lies in FinTech, with Germany being a top global FinTech hub. The industry emphasizes adoption and cooperation, with over EUR 3.76 billion invested in German FinTech in 2021. Germany is highly attractive to foreign FinTech companies, with more than 1000 international FinTech companies operating in the country (GTAI, 2022b). Ideal conditions for private equity investors attract approximately 450 associated companies investing in over 1000 German companies annually. In 2022, private equity investment reached EUR 13.9 billion, supporting various sectors (German Private Equity and Venture Capital Association, 2023).

Germany is undergoing a complex structural change that is impacting every aspect of society and the economic system. Addressing the climate crisis and consistently meeting the United Nations' sustainability goals are imperative tasks that cannot be delayed. With a noticeable shortage of skilled labor, particularly in highly specialized fields, the ability to attract and retain well-trained employees stands out as a crucial challenge for companies striving for success. As a result, many companies' sustainability focus extends beyond environmental considerations (Bannier, 2023). Instead, they are emphasizing traditional social issues like providing opportunities for employee training, ensuring healthcare benefits, and fostering a healthy work-life balance.

Moreover, certain aspects are gaining regulatory importance for companies, including robust supply chain control and the cultivation and diversity of the upcoming generation of leaders (Bannier, 2023; Federal Minsitry for Economic Affairs and Climate Action, 2022). The COVID19 pandemic and geopolitical events, like Putin's war of aggression, challenge established multilateral approaches and affect technology and industry (Zwick & Jeromin, 2023). Against this background, a sustainable financial system is essential for swift and targeted economic transformation, involving significant investments in establishing and expanding sustainable business models and realigning sectors. Substantial and urgent investment is required whether pursuing the Paris 1.5° target or advancing the international sustainability agenda. Financial market players are pivotal in driving this transformation by offering targeted financing to activities contributing to climate and other sustainability goals (Zwick & Jeromin, 2023).

The recognition of sustainable finance in Germany faced a historic delay, but positive developments have occurred. Establishing the Sustainable Finance Advisory Committee in 2019 marked a turning point, submitting 31 recommendations in its final report in February 2021. The previous German government's coalition agreement expressed optimism about making Germany a leading sustainable finance hub, emphasizing the importance of developing resilient economic structures. Timely action is crucial in the face of current challenges, and there is hope that Germany's commitment to sustainable finance will influence discussions in Brussels (Zwick & Jeromin, 2023).

6.2 The German Sustainable Finance Strategy

Given that sustainability risks involve investment risks, the German government aims to prioritize the stability of the financial market as a critical goal of its SF strategy. Losing asset value, increasing production costs, or facing higher taxes are among the consequences of implementing new environmental standards. Against this background, the financial market players must consider these risks first and foremost for their benefit and the advantage of their clientele, investors, or end users, and then in terms of the global impact on people and the environment. Germany needs an estimated €190 billion Euro annually until 2045 for the comprehensive transition of all economic sectors to climate neutrality (Kreditanstalt für Wiederaufbau, KfW Research, 2021). In 2020, about 36% of globally managed assets, totaling over $35 trillion dollars were categorized as sustainable, with Europe leading at $9 trillion dollars (Global Sustainable Investment Alliance, GSIA, 2020). Germany significantly contributed €2.2 trillion Euro in responsible investments in 2021 (Forum Nachhaltige Geldanlagen, FNG, 2022). Furthermore, there is a growing involvement of governments, international organizations, regulators, and central banks in sustainable finance (Bundesanstalt für Finanzdienstleistungsaufsicht, BaFin, 2019; European Commission, EC, 2022).

On February 25, 2019, the State Secretaries' Committee for Sustainable Development tasked the German government with creating a Sustainable Finance Strategy. The goal was to establish Germany as a prominent sustainable finance hub, focusing on the role of financial markets. To mainstream SF, Germany's national SF strategy has set for itself the following goals (German Federal Government, 2021):

1. Promote sustainable finance at the European and international level,
2. Seize opportunities, finance the transformation, consolidate the sustainability impact,
3. Improve the financial industry's risk management in a targeted way and ensure financial market stability,
4. Advance Germany as a center for sustainable finance and build expertise,
5. Establish the German government as a role model for sustainable finance within the financial system.

Consequently, the German government formed the Sustainable Finance Committee on June 6, 2019. Comprising members and observers from diverse sectors such as finance, the real economy, academia, and civil society, the committee gathers expertise from varied perspectives. The committee is crucial in guiding the government's Sustainable Finance Strategy, facilitating discussions among diverse stakeholders, and providing insights for national and European sustainability discussions (German Federal Government, 2021). In its February 25, 2021 report, "Shifting the Trillions – A Sustainable Financial System for the Great Transformation," the Committee outlined key recommendations spanning sustainability-focused policy frameworks, enhanced company reporting, research emphasis, promotion of sustainable financial products, and support for institutional consolidation. These recommendations, geared toward fostering a sustainable financial system, align with the German government's strategy, which includes specific implementation proposals based on the Committee's priorities (German Federal Government, 2021)

6.3 Implementation

We now detail the implementation of Germany's SF strategy, including EU Taxonomy adoption, content-related challenges, practical applications, and disclosure.

EU Taxonomy Adoption in Germany

The EU Taxonomy was introduced in 2020 to clarify environmentally sustainable economic activities for investors and companies. It allows companies to improve sustainability discussions, measure performance, and boost product portfolio credibility. This is especially relevant for businesses within the Taxonomy's scope, aiding strategic decision-making. However, the Sustainable Finance Advisory Committee (SFB) sees companies facing structural and content-related challenges in the practical initial application of the Taxonomy Regulation.

Structural Challenges

Companies and auditors are grappling with time-frame challenges in implementing the Taxonomy Regulation, facing short notice on new requirements and guidelines. The lack of planning time delays establishing reporting processes and internal control systems. A proposed solution emphasizes the need for increased transparency in the Commission's planned publications. Better preparation and implementation would require longer application periods (Sustainable Finance Beirat, SFB, 2023).

Legal ambiguities within the Taxonomy Regulation also pose issues, and the absence of a reliable contact point for clarification adds complexity. It is suggested that a contact point be established at the Commission to provide binding answers promptly, address interpretation questions, and ensure consistency across EU countries. Further support, such as clear implementation guidelines and tools, is also advocated (SFB, 2023).

Internationally active companies are also encountering difficulties due to the EU-centric focus of reporting requirements, potentially leading to competitive disadvantages. Moreover, references to EU requirements and directives in the Taxonomy Regulation present clarity issues and are unclear for activities outside the EU. Advancing international harmonization efforts should help create a level playing field while maintaining the EU's ambitious standards (SFB, 2023). Reviewing references for clarity and creating equivalence tables, mapping tables for criteria comparison, and a "common ground framework" for international alignment is also suggested (see Chap. 3).

Finance companies face challenges in calculating the Green Asset Ratio (GAR) due to conflicting goals, particularly about SMEs. In this regard, resolving conflicts to allow financial market participants to represent climate-friendly assets coherently is necessary. For instance, financial market participants could use information on taxonomy eligibility provided voluntarily by Special Purpose Vehicles (SPVs), with this option strictly limited to SPVs (SFB, 2023).

Content-Related Challenges

The challenges within the Taxonomy Regulation include unclear definitions, particularly for terms like "essential for society" and "best-performing alternative." Proposed solutions involve compiling clear definitions for these terms. Inconsistencies arise in economic activities, especially regarding enabling activities and differing treatment along the value chain. It is recommended to establish a framework for enabling activities, re-examine technical screening criteria, and ensure consistency in value chain interfaces. Similarly, the lack of clarity in the "Do No Significant Harm" (DNSH) criteria, with imprecise definitions and references to external documents, requires a systematic approach and clear rules for references to EU regulations. Ambiguities concerning minimum safeguards and references to non-binding documents also pose challenges. Clarifying compliance at the entity level, adjusting relevant regulations, and specifying the entity-level application of minimum safeguards might help mitigate the issue. The incomplete catalog of economic activities hampers reporting companies' ability to provide a full picture of their sustainable activities. Solutions include expanding the list to cover activities aligned with environmental taxonomy goals, introducing taxonomies for green transition and social activities, and ensuring proportionate expansions. Unrealistic expectations leading to high taxonomy alignment ratios call for emphasizing sustainability standards and communicating realistic expectations, especially for turnover Key Performance Indicators (KPIs) (SFB, 2023).

Practical Initial Application

Reporting companies must establish new processes and internal control systems (ICS) to address the challenges of implementing the Taxonomy Regulation to ensure compliance. The lack of granular internal data in financial reporting systems necessitates extensive adjustments to IT infrastructures, leading to a reliance on manual activities until these adjustments are made. Adequate documentation is crucial for auditors to review sustainability reporting effectively, emphasizing the importance of a robust ICS (SFB, 2023).

Qualified staff is essential due to the intricate nature of the Taxonomy Regulation, requiring expertise in integrated thinking and sustainability concepts. However, the current shortage of skilled workers in sustainability may lead companies to seek external consulting services. To mitigate this, comprehensive skills development through training and education programs is recommended, with support from the German government and the EU (SFB, 2023).

Insufficient data availability poses additional challenges, especially for companies integrated within value chains. External information collection difficulties, particularly for life-cycle assessments (LCAs), highlight the need for reliable data sources. Proposed solutions include EU guidance on data sources, allowances for equivalent information in the transitional period, and the expedited establishment of the European Single Access Point (ESAP). Creating public LCA databases and addressing data availability issues for regional or national comparative data are also suggested.

The challenges faced by financial companies lie in determining taxonomy key performance indicators (KPIs) based on counterparty information. Relying on counterparty reporting for loans with a known use of proceeds lacks clarity in terms of re-examination and may result in significant costs. It is emphasized that the obligation to provide documentation rests with the client, relieving banks of redundant verification processes. If appropriate documentation is unavailable, the taxonomy alignment review by the credit institution should conclude negatively (SFB, 2023).

Disclosure

A second building block of the EU Sustainable Finance Framework is disclosure. The European Commission introduced stringent disclosure requirements for nonfinancial and financial institutions on 5 January 2023 by enforcing the Corporate Sustainability Reporting Directive (CSRD) (EC, 2023). Only public interest entities must incorporate the EU Taxonomy in their reporting. The CSRD will require large companies and listed small and medium-sized companies operating in the EU to disclose sustainability data. The CSRD will replace the Non-Financial Reporting Directive (NFRD) by the end of 2024, thus affecting annual reports for 2025 (EC, 2023). The number of companies reporting under the EU Taxonomy currently includes 11,700 large companies with over 500 employees (EC, 2023). The disclosure obligations for companies affected by the EU Taxonomy differ from the disclosure obligations of financial institutions (EU Taxonomy Info, 2023). Companies and financial institutions must disclose the extent of EU Taxonomy-eligible economic activities and the proportion of EU Taxonomy-aligned economic activities since the EU Taxonomy came into force (EU Taxonomy Info, 2023). By increasing the disclosure of relevant information and signaling that companies are following an EU Taxonomy-aligned strategy, attracting more sustainability-oriented investors and gaining a potential competitive advantage might be possible (EU Taxonomy Info, 2023).

The European Commission has created a science-based uniform classification system for (environmentally) sustainable economic activities with the EU Taxonomy. The aim is for companies—building on the reporting obligations of the Taxonomy Regulation—to report on the impact of sustainability aspects on their business activities and their impact on man and the environment in a more detailed and standardized manner, going beyond specific balance sheet. By extending the scope of application to smaller companies, the aim is also to ensure that more companies than before report following the directive's requirements (Gaedicke & Willert, 2023).

The CSR Directive Implementation Act in Germany, responding to the European Non-Financial Reporting Directive (NFRD 2014/95/EU), mandated large listed companies to include a non-financial statement in their management reports starting from the fiscal year 2017. While the directive aimed to standardize sustainability reporting across the EU, implementing national laws has led to significant variations, affecting both the types of companies involved and the content and auditing of reports (Bannier, 2023).

Public Interest Entities (PIEs), encompassing banks, insurance companies, and market-related corporations with over 500 employees, are subject to this reporting obligation. However, the EU directive's narrow criteria resulted in around 2000 affected companies, primarily comprising large publicly listed companies in the real economy. Member states, including Germany, can define the scope further. For instance, some countries like Denmark and Sweden set a lower employee threshold for market-oriented companies (Bannier, 2023).

As per the EU directive, non-financial reports must cover environmental, social, and employee aspects, human rights, and anti-corruption measures. However, the specific requirements and additional topics vary across countries. In Germany, no additional quantitative disclosures are mandated. Companies can integrate the nonfinancial report into the management report or adopt a separate format. Germany requires publication within the management report or, at the latest, 4 months after the balance sheet date. While the board is responsible for preparing the non-financial statement, and the supervisory board reviews its content, there is no mandatory external auditor for substantive auditing, unlike some EU countries like Denmark. Many companies voluntarily engage audit firms for a limited assurance audit of their non-financial statements. In 2020, 37% of the 160 companies in Germany's DAX family opted for such an audit. The variations in the implementation of the directive highlight the challenges in achieving uniformity across EU member states.

Due to impending changes, German companies are currently grappling with uncertainties in their information and communication policies. The Taxonomy Regulation, linked to the CSRD, mandates new reporting obligations in sustainability, effective since January 1, 2022, but lacks implementation details. This ambiguity in regulatory interpretation poses challenges for companies, especially as they face inquiries from regulators, investors, customers, and employees on sustainability. Establishing a comprehensive reporting system incurs high costs, prompting a need to reflect on past experiences with sustainability reporting abroad and anticipate potential effects on reporting companies, emphasizing what makes these reports effective.

CSRD-affected firms should swiftly gather and verify data, drawing from prior sustainability reporting experiences. A robust materiality analysis is recommended to streamline the process, especially for SMEs. This involves pinpointing key stakeholders and vital sustainability areas, shaping reporting priorities, and influencing strategy, risk management, and governance for enhanced overall business sustainability.

6.4 Sustainable Finance Key Players in Germany

In Germany, the sustainable finance regulatory landscape is overseen by several key authorities, each playing a critical role in shaping and implementing policies to drive sustainable financial practices. The Federal Financial Supervisory Authority (BaFin) is the primary regulatory body overseeing financial markets, including banks, insurers, and securities trading. Alongside BaFin, the Kreditanstalt für Wiederaufbau (KfW Bank) promotes sustainable development through its financing and investment activities, particularly in infrastructure and environmental projects. Furthermore, the Federal Ministry of Finance plays a central role in formulating and implementing policies related to sustainable finance, providing strategic guidance and legislative frameworks to ensure the integration of sustainability considerations within the financial sector. The following subsections will detail each of the key activities undertaken by these authorities.

BaFin

BaFin, the German Federal Financial Supervisory Authority, recognizes the significant impact of ESG factors (environmental, social, and governance) on the financial sector. It stresses the need for supervised companies to manage associated risks effectively. At the end of 2019, BaFin formulated its expectations of supervised financial institutions in the "Guidance Notice on Dealing with Sustainability Risks" (BaFin, 2019). The fact sheet was presented as a "compendium of non-binding codes of conduct" (BaFin, 2019). However, this non-binding nature only relates to the specific structure. There is no question that companies must integrate sustainability risks into their strategies, business organization, and risk management (Gaedicke & Willert, 2023).

To empower investors to make decisions aligned with sustainability preferences, BaFin emphasizes the need for comprehensive and understandable information. Transparency is a critical focus that enables consumers to make independent and informed decisions. BaFin's medium-term objectives include integrating sustainability into supervisory activities, mainly focusing on environmental aspects, notably climate change (BaFin, 2023).

At the European level, BaFin is actively driving efforts to enhance the quantification, measurement, and management of sustainability-related risks in collaboration with the three European Supervisory Authorities (ESAs) (German Federal Government, 2021). BaFin also holds a position on the steering committee of the Network for Greening the Financial System (NGFS) and participates in expert groups addressing climate-related financial market risks led by standard-setting bodies such as the Basel Committee on Banking Supervision and the Financial Stability Board (German Federal Government, 2021). As a supervisory authority, BaFin recognizes the pivotal role of targeted capital allocation in realizing a climate-neutral and sustainable economy. Various sustainable finance initiatives at both European and national levels are focused on channeling private funds into activities that promote sustainability (German Federal Government, 2021).

In collaboration with the Federal Ministry of Finance and the Deutsche Bundesbank, BaFin is committed to enhancing global dialogue and supporting initiatives like the Coalition of Finance Ministers for Climate Action (CFMCA) and the Network of Central Banks and NGFS. This includes organizing follow-up events, such as the one after the joint CFMCA and NGFS workshop on risk management in November 2020, hosted by the Federal Ministry of Finance and BaFin.

Key areas of BaFin's current focus involve risk-oriented and practical regulation, improving data on financial climate risks, managing climate-related financial risks within existing categories, preventing greenwashing, and fostering an open dialogue for knowledge exchange. The authority emphasizes proportionate and risk-based supervisory priorities, utilizing existing tools for effective implementation. BaFin actively monitors how supervised companies disclose sustainability risks, emphasizing the importance of reliable data, especially with evolving disclosure requirements. It recognizes environmental risks as specific drivers and encourages continuous efforts by companies to identify, measure, and manage these risks. Preventing greenwashing and presenting misleading sustainability information is a significant concern for BaFin. Increased transparency through specific disclosure and reporting requirements is crucial to align investor decisions with sustainability preferences. BaFin utilizes its supervisory tools to prevent greenwashing, particularly in product and market supervision (BaFin, 2023).

The complexity of sustainable finance necessitates an open dialogue, and BaFin actively engages in discussions with stakeholders, contributing to international initiatives and supporting the development of foreign supervisory systems. It participates in the Sustainable Finance Advisory Committee of the Federal Government, contributing to ESG discussions led by coordinating authorities. It places great importance on knowledge exchange and maintaining an open dialogue with experts in navigating the intricate landscape of sustainable finance.

KfW Bank

KfW, as a global promotional bank representing the German Federal Government and the Federal States, is actively embracing the challenge of steering toward a sustainable society. Recognizing the pivotal nature of this transformation, KfW aligns its objectives with the international community's sustainability goals, those of the European Union, and the specific targets outlined by the Federal Republic of Germany (KfW, 2023). In a bid to play a pioneering role in the financial sector, KfW is dedicated to substantially contributing to the UN Sustainable Development Goals and the Paris Climate Agreement. Central to this commitment is a strategic shift in the bank's steering, facilitated by the gradual implementation of a comprehensive Sustainable Finance concept across the entire KfW Group. KfW's ambitions and sustainable finance concept are endorsed by the German government's inaugural strategy for sustainable financing activities. This strategy supports KfW in evolving into a transformative promotional bank that aligns with the Climate Protection Programme 2030, paving the way for a sustainable and greenhouse gas-neutral future (KfW, 2023).

Since January 2021, all divisions within KfW have consistently implemented the Sustainable Finance concept. Notable changes in the strategic target system underscore the primary goal of fostering transformation. Two additional goals, namely "KfW's financing activities' SDG-contribution" and "Paris compatibility of KfW's financing activities," emphasize the commitment to sustainability. Key mechanisms within KfW Group's Sustainable Finance framework include the mapping of promotional activities based on the UN Sustainable Development Goals (SDG-Mapping), comprehensive impact management (including greenhouse gas accounting), and the introduction of sector guidelines to ensure the compatibility of financing activities with the objectives outlined in the Paris Climate Agreement.

Federal Ministry of Finance

In the area of climate policy, a core aim within the Finance Track is to ensure that financial institutions focus more on sustainability in their decision-making (sustainable finance). For this reason, the G7 under Germany's presidency not only welcomed the work of the International Sustainability Standard Board (ISSB) but also took concrete steps to ensure close cooperation between the ISSB and EU standard-setters as well as a stronger focus on the interests of emerging and developing countries. The G7 also agreed to carry out the necessary preparations to apply the ISSB Baseline. In this way, the ISSB has been given a strong mandate as the global standard-setter for sustainability related corporate reporting (Federal Ministry of Finance, 2023).

Moreover, the Federal Ministry of Finance has been conducting a series of sustainable finance workshops, including sessions at German embassies, with the involvement of the Federal Foreign Office, the Federal Ministry for the Environment, Nature Conservation and Nuclear Safety, the Federal Ministry for Economic Cooperation and Development, and the Federal Ministry for Economic Affairs and Energy (German Federal Government, 2021). These workshops address topics like sustainable corporate governance in both the real economy and financial sector, alongside issues of sustainability, human rights, environmental risk management, transparency, financing instruments (e.g., green or sustainability bonds), and financial market regulation.

6.5 Conclusion

Germany's financial sector, boasting global recognition, plays a pivotal role in the nation's economic landscape. As the country emphasizes sustainability, the financial industry is undergoing significant transformations and adapting to global standards (Path time line in Chap. 11, Fig. 11.2). The focus on sustainable finance aligns with Germany's commitment to environmental and social considerations. However, challenges persist in implementing frameworks such as the EU Taxonomy, as seen in the structural, content-related, and practical challenges.

Regulators and key players like BaFin, KfW Bank, and the Federal Ministry of Finance actively contribute to the sustainability agenda. BaFin's emphasis on transparency and risk management, KfW Bank's strategic shift towards sustainable finance, and the Federal Ministry of Finance's workshops on sustainable finance underscore a collective effort toward achieving environmental and social goals. Despite progress, there is a need for continued collaboration and adaptation. The complexity of sustainable finance and evolving regulations require a proactive approach. Enhancing transparency, addressing challenges associated with the EU Taxonomy, and fostering international cooperation will be essential. Additionally, investing in developing skills related to sustainable finance and providing support to companies navigating these changes will contribute to a smoother transition.

References

Bannier, C. (2023). Nachhaltigkeitsberichterstattung – Aktuelle Herausforderungen und Chancen für Großunternehmen und Mittelständler. In *Mit Sustainable Finance die Transformation dynamisieren* (pp. 159–170). Springer Fachmedien Wiesbaden.

Bundesanstalt für Finanzdienstleistungsaufsicht (Federal Financial Supervisory Authority - BaFin). (2019). *Merkblatt zum Umgang mit Nachhaltigkeitsrisiken*. Retrieved December 10, 2023, from https://www.bafin.de/SharedDocs/Downloads/DE/Merkblatt/dl_mb_Nachhaltigkeitsrisiken.html

Bundesanstalt für Finanzdienstleistungsaufsicht (Federal Financial Supervisory Authority). (2023). *BaFin's sustainable finance strategy*. https://www.bafin.de/EN/DieBaFin/Sustainable_Finance_Strategie/SF_Strategie_node_en.htm
EU Taxonomy Info. (2023). *Application of the EU Taxonomy for companies*. https://52eu-taxonomy.info/info/eu-taxonomy-for-companies
European Commission. (2022). *Platform recommendations on data and usability*. https://finance.ec.europa.eu/system/files/2022-10/221011-sustainable-finance-platform-finance-report-usability_en_1.pdf?showiframe=true
European Commission. (2023). *Corporate sustainability reporting*. https://finance.ec.europa.eu/capital-markets-union-and-financial-markets/company-reporting-and-auditing/company-reporting/corporate-sustainability-reporting_en
Federal Minsitry for Economic Affairs and Climate Action. (2022). *Federal ministry for economic affairs and climate action (BMWK) public relations*. Retrieved January 7, 2024, from Bmwk.de. https://www.bmwk.de/Redaktion/EN/Publikationen/Wirtschaft/annual-economic-report-2022.pdf?__blob=publicationFile&v=2
Federal Statistical Office. (2024). Current population of Germany. Federal Statistical Office. https://www.destatis.de/EN/Themes/Society-Environment/Population/Current-Population/_node.html
Forum Nachhaltige Geldanlagen. (2022). *Marktbericht Nachhaltige Geldanlagen 2022*. https://fng-marktbericht.org/fileadmin/Marktbericht/2022/FNG-Marktbericht_NG_2022-online.pdf
Gaedicke, M., & Willert, M. (2023). *Die Rolle der Versicherer in der nachhaltigen Transformation*. In *Mit Sustainable Finance die Transformation dynamisieren* (pp. 65–80). Springer Fachmedien Wiesbaden.
German Federal Government. (2021). German Sustainable Finance Strategy. Retrieved January 7, 2024, from Bundesfinanzministerium.de. https://www.bundesfinanzministerium.de/Content/DE/Downloads/Broschueren_Bestellservice/deutsche-sustainable-financestrategie.pdf.
German Private Equity and Venture Capital Association. (2023). https://www.bvkap.de/statistics/statistics-germany
Germany Trade and Invest (GTAI). (2022a). *The Finance Sector in Germany*. Retrieved January 7, 2024, from Gtai.de website: https://www.gtai.de/en/invest/service/publications/finance-in-germany-863564
Germany Trade and Invest (GTAI). (2022b). *Fintech in Germany*. Retrieved January 7, 2024, from Gtai.de website: https://www.gtai.de/resource/blob/159528/d297a4985e550343a6a0d-c08fa046124/20220922_FactSheet_FinTech_WEB.pdf
Global Sustainable Investment Alliance (GSIA). (2020). *Global Sustainable Investment Review 2020*.
International Trade Administration. (2023). *Germany Country Commercial Guide*. https://www.trade.gov/country-commercial-guides/germany-market-overview?section-nav=1956
KfW. (2023). *Sustainable finance*. https://www.kfw.de/nachhaltigkeit/About-KfW/Sustainability/Strategie-Management/Sustainable-Finance/
KfW Research. (2021). *5 Bio. EUR klimafreundlich investieren – eine leistbare Herausforderung, Fokus Volkswirtschaft, 350*. https://www.kfw.de/PDF/Download-Center/Konzernthemen/Research/PDF-Dokumente-Fokus-Volkswirtschaft/Fokus-2021/Fokus-Nr.-350-Oktober-2021-Investitionsbedarfe-Klimaneutralitaet.pdf
Sustainable Finance Beirat (Sustainable Finance Advisory Committee of the German Federal Government). (2023). *The EU Taxonomy: Implementation challenges and proposed solutions*. https://sustainable-finance-beirat.de/wp-content/uploads/2023/03/SFB_The-EU-Taxonomy_implementation-challenges-and-proposed-solutions.pdf
Zwick, Y., & Jeromin, K. (Eds.). (2023). *Mit Sustainable Finance die Transformation dynamisieren: Wie Finanzwirtschaft nachhaltiges Wirtschaften ermöglicht* (1st ed.). Springer Gabler.

Saadat Hussain is a research associate at the Center for Social Investment, Ruprecht-Karls-Universität Heidelberg, Germany.

Chapter 7
Estonia: Sustainable Finance in a Progressive Nation

Kristiina Kerge

Abbreviations

CSRD	Corporate Sustainability Reporting Directive
EIBIS	The European Investment Bank's Investment Survey
EsBA	The Estonian Banking Association
ESG	Environment, Social, and Governance
ESRS	European Sustainability Reporting Standards
EU	European Union
GDP	Gross domestic product
GHG	Greenhouse gas
GRI	Global Reporting Initiative
ICT	Information and communications technology
ID	Identity Document
km^2	Square kilometer
kt	1000 ton
LULUCF	Land use, land-use change, and forestry
NGFS	Network for Greening the Financial System
PPPs	Public-Private Partnerships
RD	Research and Development
RRF	Recovery and Resilience Facility
TSI	Technical Support Instrument

K. Kerge (✉)
Estonian University of Life Sciences, Tartu, Estonia
e-mail: kristiina.kerge@zwee.org

R. Calipha et al. (eds.), *Sustainable Finance Regulation in the European Union*, Sustainable Finance, https://doi.org/10.1007/978-3-031-88392-7_7

7.1 Introduction

Estonia stands as a testament to the power of local governance and civic engagement in the heart of the European Union. Despite its modest size, both geographically at 45,339 km^2 and demographically with a population of 1,365,884 (in year 2023) his nation distinguishes itself through a highly decentralized approach to governance. Across its 79 local government units, Estonia entrusts significant autonomy to local authorities, empowering them to address all matters of local life independently from the state. This autonomy includes the governance and administrative aspects and extends to financial independence, with local governments wielding the authority to manage their budgets and levy taxes.

Central to Estonia's model of local governance is the council, a representative body elected by the citizens every 4 years. In this context, participatory budgeting emerges as a pivotal mechanism for democratizing financial decision-making. In 2023, 50 out of 79 local government units have embraced participatory budgeting, allocating funds ranging from 5000 to 1,000,000 euros to projects directly chosen by their communities. This initiative not only fosters greater community involvement and transparency in allocating public funds but also underscores Estonia's commitment to inclusive and sustainable development. The origins of participatory budgeting in Estonia are traced back to 2011, spearheaded by the e-State Academy's efforts to promote this innovative practice. Tartu, Estonia's second-largest city, led the way in 2013 by implementing its first inclusive budget, setting a precedent for other cities and municipalities. This forward-thinking approach to budgeting has since become a cornerstone of Estonia's democratic governance, allowing citizens to shape their local environment and community projects directly. Through participatory budgeting, Estonia exemplifies how small nations can lead significant advancements in democratic practices and sustainable development, providing a model for others to emulate.

Environmental Characteristics

Estonia's journey toward the ambitious goals of the European Green Deal, particularly achieving climate neutrality by 2050, is marked by its commitment to fostering a sustainable society where solidarity ensures a healthy living environment for all. However, this path has challenges, especially within the energy sector. Historically reliant on fossil fuels, Estonia's energy production has been a significant source of CO2 emissions, with the industry accounting for nearly 70% of national emissions in 2017, placing Estonia among Europe's highest per-capita CO2 emitters. A noticeable shift occurred in 2019 when oil shale electricity production declined, influenced by escalating CO2 prices within the EU and a concerted push towards green energy. Despite this progress, the contemplation of up to four new shale oil plants and a pre-refinery presents a complex scenario for Estonia's environmental policies

(Kliimamuutused.ee, n.d.). Moreover, Estonia's land use, land-use change, and forestry (LULUCF) sector have witnessed a critical transition from a net absorber to a net emitter of CO2 equivalents. This alarming reversal, as detailed in Annex 6 of a report, from net removals of 4835 kt CO2-equivalent in 2010 to net emissions of 1297 kt CO2-equivalent in 2020, is mainly due to a marked decrease in forest carbon sequestration capabilities. Addressing this issue urgently, Estonia recognizes the importance of enhancing forest management and adopting a robust national forest development plan to restore this vital absorption capacity (European Commission, 2023a, 2023b).

The Bank of Estonia highlights that Estonian firms are primarily motivated by the desire to bolster their reputation and the necessity to mitigate high energy, water, and materials costs. This mirrors the critical drivers for environmental innovation observed across the EU. Remarkably, one of ten Estonian firms engages in environmental innovation driven by these factors (Eesti Pank, 2023).

Estonians generally hold positive views towards environmental conservation, with a significant majority understanding the finite nature of our planet's resources. In 2022, 83% agreed that "The Earth is like a spaceship with little space and limited resources." Approximately 80% consider themselves environmentally conscious. However, there remains an evident gap between this awareness and a comprehensive understanding of how individual behaviors impact the environment. Closing this gap is essential for translating environmental consciousness into practical and sustainable action (Ministry of the Environment (MoE), 2022).

Economic Characteristics

Estonia's economic landscape is deeply intertwined with the European Union, with intra-EU trade comprising 66% of its exports, notably including Finland at 16%, Sweden at 10%, and Latvia at 9%. Notably, 8% of exports reach the United States, demonstrating Estonia's significant economic connections within and beyond the EU. Since regaining independence in 1991, there has been a steadfast commitment across all administrations to align with the Western European economic sphere. This vision has transcended the country's frequent political shifts and has been foundational to its economic strategy (Staehr, 2004).

Transitioning from its post-Soviet economic structures, Estonia has established itself as a digital pioneer and an incubator of high-value tech startups, often called a "unicorn factory." This transformation is attributed to deliberate government policies, strategic investments in education and digital infrastructure, and a robust entrepreneurial ethos. Key initiatives from the early 2000s, such as the development of high-speed internet, the implementation of a digital ID system, and the launch of the "e-Estonia" initiative, have been instrumental in fostering a digital-forward economy, promoting digital service adoption across public and private sectors (Kekkonen et al., 2023).

Despite these advancements, Estonia faces the challenge of harmonizing its economic progression with sustainable practices. The nationwide push towards sustainability has largely been propelled by EU-wide regulations rather than being driven by solid internal policies. Estonian businesses predominantly perceive the green transition as a challenge, with concerns centered around financial constraints, including rising costs and limited access to green financing. Additionally, the lack of robust policy support, tepid consumer demand for green products, and the uncertainties surrounding regulatory frameworks have compounded the complexities of this transition (Kekkonen et al., 2023). As Estonia navigates this pivotal juncture, fostering an environment where sustainability is viewed not just as an obligation but as an economic opportunity may be vital to unlocking the next phase of the country's economic evolution.

7.2 Key Financial Institutions

The Estonian financial sector has demonstrated a robust expansion over the last decade and is considered medium-sized when assessed on an international scale. The financial assets of this sector amounted to approximately 240% of the nation's GDP at the close of 2022 when viewed in unconsolidated terms. This figure represents a growth of about one-third over the preceding 10 years, positioning it slightly below the European Union's median. Reflecting on the financial sector's contribution to Estonia's economy, there has been an appreciable increase in its role in generating value added; it contributed 4.6% of the total value-added in 2022, rising from 4% a decade ago, which is marginally higher than the EU average of 4.3% for the same year (The Structure of the Estonian Financial Sector, 2023).

Central to Estonia's financial system are several pivotal institutions, including Eesti Pank, the Financial Supervision Authority, and the Estonian Banking Association. Eesti Pank, as the central bank, took a significant step towards sustainability by joining the Network of Central Banks and Supervisors for Greening the Financial System (NGFS) in 2020. This initiative, launched in December 2017 by eight founding members, has seen its membership swell to 121 by October 2022, encapsulating all euro area central banks and 19 observers. The NGFS operates as a voluntary consensus-based forum, promoting the integration of environmental considerations into the fabric of financial governance (Eesti Pank, n.d.-a, n.d.-b).

In a landmark move, Eesti Pank's strategy for autumn 2021 underscored sustainable development and climate issues, highlighting a strategic pivot towards integrating these critical considerations into its core functions. As a backbone of the Estonian financial system, Eesti Pank plays a crucial role in shaping the financial sector's policy landscape, participating in legislative development, and framing crisis management protocols. This strategic emphasis on sustainable development represents a pivotal alignment with broader global trends toward greening the financial system. It underscores Estonia's commitment to addressing climate-related risks as part of its financial policy agenda (Eesti Pank, 2023).

The Finantsinspektsioon, serving as Estonia's financial supervisory and crisis resolution authority, operates with a distinct blend of autonomy and state-backed authority, ensuring its independent decision-making process. In 2022, it took a significant stride toward sustainable finance by releasing the Sustainability Investment Roadmap. This pivotal document delineates supervisory expectations for financial market companies, guiding their sustainable activities and aligning with global trends towards environmentally conscious investments (Finantsinspektsioon, 2022). Complementing its regulatory responsibilities, Finantsinspektsioon is also a proponent of financial technology innovation. This commitment was formalized with the establishment of 'The Innovation Center' on June 17, 2021, following the enactment of a legislative amendment to the Financial Supervision Authority Act. This amendment empowered Finantsinspektsioon to foster financial technology development within its regulatory purview and integrated this mandate into the institution's strategic vision. The Innovation Center's inception aligns with Finantsinspektsioon's broader mission to protect investors and consumers, contributing to the financial sector's stability, reliability, transparency, and efficiency. Moreover, it underscores the authority's dedication to mitigating systemic risks and safeguarding the financial industry against misuse of illicit activities (Innovation Center report, 2022). Integrating the provided section into the previous paragraph with clarity and additional context, we get the following improved paragraph:

The Estonian Banking Association (EsBA), a key pillar in Estonia's financial domain, comprises 10 member banks that collectively represent approximately 98% of the banking sector's assets. These member banks span from significant subsidiaries and affiliates of Scandinavian banking groups to Estonian-owned institutions, reflecting a financial landscape where Nordic financial groups have substantial holdings. Notably, the largest banks—Swedbank, SEB, Luminor, and LHV—are part of these groups, which frequently also possess insurance companies, fund managers, and lease companies. In contrast, a growing segment of the financial sector, including all payment institutions, savings and loan associations, a majority of fund managers, as well as private equity and venture capital funds, creditors, and investment firms not linked to banks, is held by Estonian entities (The Structure of the Estonian Financial Sector, 2023).

In Estonia, banks offer a spectrum of services, from daily banking needs to advanced wealth management, and they are the primary providers of leasing and factoring services The EsBA's commitment to sustainability is evident in its establishment of a dedicated Committee on Sustainable Banking, which aims to ensure that the banking sector is equipped to introduce financial products and services that facilitate Estonia's sustainable growth. The key financial institutes, including these banks, are instrumental in shaping the country's approach to sustainable finance, developing strategies, and setting environmental goals within their organizations. These institutions also play a crucial role as stakeholders in policy formation for the financial sector.

Currently, implementing sustainable financing practices in Estonia is still nascent, with no extensive research available to detail the unique methods and obstacles the country encounters in advancing sustainable finance. The EsBA's role is thus pivotal, actively participating in sustainability-related activities, steering discussions on funding for necessary investments, and ensuring the readiness of banks to offer new, sustainable financial products to consumers. A comprehensive list of Estonian banks spearheading these efforts can be found on the EsBA's website, which provides a clearer picture of those leading the charge towards a sustainable financial future. To enhance the financial sector's sustainability, these institutions are not only driving the adoption of green policies. Still, they are also pivotal in designing the broader regulatory and operational landscape that will define Estonia's sustainable finance practices in the future.

7.3 The Estonian Regulatory Framework

The "Estonia 2035" strategy represents a pivotal national vision articulated through a comprehensive and inclusive process that lasted 2 years and engaged nearly 17,000 participants in meaningful dialogue (Government of Estonia, 2022). The outcome is a set of five value-based strategic goals, which form the cornerstone of the country's long-term priorities. These goals are not only central to the strategic choices that Estonia will make but also serve as guiding principles for all strategic development documents. Furthermore, these objectives are carefully integrated into the state budget strategy and the government's action program, ensuring that Estonia's trajectory aligns with domestic needs and international directives (Government of Estonia, 2023). The strategic goals outlined in the "Estonia 2035" strategy, are operational through a multi-faceted approach, where actionable policies and initiatives are carefully crafted and implemented to drive the nation towards its long-term visions. Specifically, for the "Economy and climate goal," Estonia has initiated a series of targeted measures that translate high-level objectives into tangible actions:

1. Innovation and RD Investment: The government incentivizes private sector investment in research and development, particularly in areas that contribute to sustainable growth. This includes tax incentives for RD activities, grants for innovation projects, and support for partnerships between businesses and research institutions.
2. Education and Workforce Development: Estonia is focusing on enhancing its educational programs to produce a skilled workforce adept at meeting the demands of a sustainable economy. This includes revising curricula to include sustainability and climate science and creating specialized programs to train researchers and engineers.
3. Digital Transformation: Leveraging its position as a leader in digital technology, Estonia is driving the digital transformation of its economy to improve efficiency, reduce waste, and lower carbon footprints. This includes digitalization of energy grids, smart city initiatives, and support for developing green tech solutions.
4. Green Financing: Estonia is working to expand green financing options to support investments in sustainable projects. This could involve the development of

green bonds, sustainability-linked loans, and other financial products that encourage environmentally responsible entrepreneurship.
5. Regulatory Reforms: The government is also reviewing and updating its regulatory frameworks to support sustainable business practices, such as setting higher environmental standards for industries, implementing carbon pricing mechanisms, and encouraging transparency in corporate sustainability reporting.
6. Public-Private Partnerships: To achieve the ambitious targets, Estonia is promoting public-private partnerships (PPPs) that focus on sustainable development. These partnerships aim to leverage the strengths of both sectors in addressing climate challenges, such as developing renewable energy projects or energy-efficient infrastructure.
7. Sustainable Infrastructure: Investment in sustainable infrastructure is another key initiative, with the government funding the development of green transportation, energy-efficient buildings, and renewable energy sources, which are all critical to reducing the country's carbon footprint.
8. International Collaboration: Estonia is actively engaging in international collaboration to both learn from global best practices in sustainable finance and climate action, and to contribute its expertise in digital innovation to the international community.

By integrating these policies and initiatives into the fabric of its economy, Estonia aims to increase the number of researchers and engineers in the private sector and foster a broader ecosystem that supports sustainable entrepreneurship and climate-conscious economic growth.

Corporate Sustainability Reporting Directive (CSRD)

Estonia is in the implementation phase to add the Corporate Sustainability Reporting Directive (CSRD) is included in national legislation. With the CSRD coming into force at the beginning of 2023, the directive mandates particularly large companies to provide comprehensive disclosures on their environmental and social impacts, risks, opportunities, and the reciprocal effects of the environment and society on the company's operations across various timeframes. Approximately 300–350 Estonian companies are expected to fall under this reporting requirement, with a third expected to utilize the consolidation exception. The consultation process for CSRD implementation has been thorough and inclusive, involving a wide array of stakeholders, including sectoral policymakers from various ministries, business organizations, professional associations, key financial institutes, sectoral interest groups, and entrepreneurs. Notably, the Estonia Financial Supervision Authority highlighted the need for a centralized knowledge hub to understand and apply sustainability reporting guidelines and standards. Country specific activities in Estonia include:

- September 2023—the government supported adopting a delegated regulation by the European Commission establishing European Sustainability Reporting Standards (ESRS).

- Starting in July 2024, Estonia will enforce the Corporate Sustainability Reporting Directive (CSRD) requirements within its accounting, auditing, and securities market legislation.
- In the first quarter of 2024, the Estonian Ministry of Finance is set to launch a dedicated website designed to assist entrepreneurs by consolidating all essential information required for compliance with the Corporate Sustainability Reporting Directive (CSRD).
- From 2023 to 2025, the Estonian Ministry of Economic Affairs and Communications provided support to the SusTool project, funded by the Interreg Baltic Sea Region program (Interreg Baltic Sea Region, n.d). This initiative aims to develop and pilot a bespoke online tool for service and ICT sector companies. The tool is designed to help these companies identify their environmental and social impacts and risks, as well as provide recommendations aligned with European Sustainability Reporting Standards (ESRS).

Implementation of the EU Sustainable Finance Taxonomy

The EU Taxonomy for Sustainable Activities, a cornerstone of the European Green Deal, became effective in July 2020. This innovative classification categorizes economic activities based on their environmental sustainability across six key objectives, including climate change mitigation, climate adaptation, sustainable use and protection of water and marine resources, transition to a circular economy, pollution prevention and control, and security and restoration of biodiversity and ecosystems (see Sect. 7.2). Estonia's commitment to these objectives is evidenced by the country's alignment with the EU's Sustainable Finance Action Plan, which stipulates what constitutes an environmentally sustainable economic activity. The taxonomy's actual value for Estonia lies in its potential to guide substantial investment flows into green initiatives, thereby enabling larger companies, banks, and other financial institutions to contribute more effectively towards meeting the environmental and climate objectives. In support of Estonia's efforts to implement necessary reforms that align with the Green Deal's ambitious climate target of carbon neutrality, the European Commission has extended assistance through the Technical Support Instrument (TSI). This EU program is designed to offer bespoke technical expertise to member states, enabling them to craft and execute critical reforms. The TSI has been instrumental in aiding the Estonian Ministry of Climate and the Ministry of Finance of Latvia in pursuing the Green Deal's objectives, harmonizing national efforts with the United Nations' Sustainable Development Goals (Technical Support Instrument, 2021).

The October 2023 publication of the Sustainable Finance Roadmap for Estonia represents a strategic advance in the country's commitment to environmental stewardship and financial innovation. The roadmap's project team has outlined a vision for establishing a Knowledge Hub in Estonia, with one of its key objectives being enhancing data availability. Recognizing that access to comprehensive sustainability-related data is crucial for expanding green lending and the adoption of green financial instruments. The data sets identified as vital to the growth of sustainable finance encompass a wide array of metrics, including energy efficiency in various sectors, overall energy consumption, water usage, the composition of vehicle fleets by sector and fuel type, waste management practices, and the actual rates of material recycling, as well as greenhouse gas emissions across different industries. The envisioned Knowledge Hub would monitor and address any shortcomings in the availability of such data, establishing clear responsibilities for bridging these gaps (KPMG Baltics, Eversheds Sutherland Bitāns, 2023). Moreover, the Roadmap acknowledges the fluid nature of sustainable finance and economic development concepts, pointing to the lack of extensive experience in applying new regulations. The Knowledge Hub is expected to lay the foundation for expert dialogue and the consensus-building necessary to identify and disseminate best practices for the implementation of sustainable finance regulations, promoting their uptake among financial and non-financial sector entities alike (KPMG Baltics, Eversheds Sutherland Bitāns, 2023).

State Mechanisms and Financial Instruments for Sustainable Transition

The State Recovery and Resilience Facility (RRF) underscores its commitment to fostering a sustainable and resilient future. The RRF, central to Estonia's plan, dedicates 59% of its funds to climate objectives, reinforcing the nation's dedication to environmental stewardship as part of its broader economic strategy. With Council approval secured on 29 October 2021, the plan received a substantial update on 16 June 2023, integrating reforms and investments to achieve the ambitious REPowerEU objectives, thereby enhancing Estonia's capacity for green and digital innovation (European Commission, 2023a, 2023b). A notable example of this strategic funding is the SmartCap Green Fund, infused with €100 million Euro from the European Union's Next Generation EU Recovery and Resilience Facility. From 2022 to 2026, this fund is set to channel investments through private venture capital funds to bolster financial capital for burgeoning green technology companies (SmartCap Green Fund, n.d.).

The EU's financial commitment to Estonia's updated recovery plan totals €953.3 million, reflecting a robust support structure for the nation's transitionary goals approved on 16.06.2023. The Estonian Business and Innovation Agency also provides a specific support mechanism for manufacturing companies looking to pivot their business models toward sustainable practices, with a fund of €8.7 million (Estonian Business and Innovation Agency, n.d.). Despite the availability of €20 million to mitigate supply chain disruptions amid the energy crisis, it's notable that sectors such as agriculture, forestry, and fishing are presently ineligible for this support, presenting a potential area for policy review and inclusive support expansion.

Complementing these funds is the EU's Just Transition Fund, a strategic financial tool to alleviate the socio-economic impacts of transitioning towards the 2030 energy and climate targets and 2050 climate neutrality goal. In Estonia, the Ministry of Regional Affairs and Agriculture is the driving force behind realizing these objectives. In Autumn 2023, the state unveiled a €17 million support mechanism for the Ida-Virumaa region, designed to aid the community in achieving climate neutrality. This support is tailored for initiatives such as fostering social entrepreneurship, raising environmental consciousness, safeguarding local cultural heritage post-industrialization, and engaging the youth in sustainable activities.

However, while the Global Reporting Initiative (GRI) standards have been widely recognized for a decade and adopted by 95% of large international firms for sustainability reporting, a 2022 study by Sustinere reveals that only 21 out of Estonia's 100 most influential companies have adopted a deliberate and comprehensive approach to sustainability disclosure (Sustinere, 2022). This discrepancy underscores the need for enhanced awareness and integration of sustainability reporting standards within the Estonian corporate sector.

These initiatives, from significant EU funding mechanisms to targeted local support programs, contribute to a tapestry of efforts shaping Estonia's green transition. As the nation continues to refine its sustainable finance framework, these mechanisms serve as a financial lifeline and a strategic guide towards a greener and more resilient economy.

7.4 Challenges in Sustainable Finance Implementation

Implementing sustainable finance in Estonia faces distinct challenges, a critical one being the limited understanding of sustainable finance and ESG criteria among non-financial sector companies needing operational and investment financing. Furthermore, there's a noticeable variance in the financial sector companies and investors' grasp of integrating sustainability principles into their funding and investment decisions. This disparity has been highlighted by the TSI project led by KPMG Baltics and Eversheds Sutherland Bitans in 2023. The European Investment Bank's Investment Survey (EIBIS) 2023 report underscores this concern, indicating that Estonian businesses are lagging in climate change initiatives (European Investment Bank, EIB, 2023). A comparative analysis with the previous year reveals a drop in Estonian companies setting and tracking their greenhouse gas (GHG) emissions targets. In 2022, 40% of Estonian firms were proactive in this regard, aligning with the EU average, but there has since been a reduction.

On a positive note, innovative platforms like eAgronom have facilitated knowledge transfer in the agricultural sector, enabling farmers to gain certification for sustainable farming, opening the door to specialized financing options. Swedbank in Estonia capitalizes on this by offering more favorable financial services to certified sustainable farmers, with eAgronom serving as a crucial partner in issuing and monitoring these certificates (Swedbank Estonia, n.d.). SEB Bank's initiative to provide loans for solar panel installations, collaborating with Explorer to ensure customers receive expert guidance, further exemplifies the proactive steps being taken. Meanwhile, LHV Bank has adopted a comprehensive approach to sustainability, offering specific products to finance green activities (LHV 2023). However, they also acknowledge the challenge of aligning sustainability with financial profitability, as investors typically prioritize returns, viewing sustainability as a supplementary benefit rather than a primary goal. The launch of the Cleantech for Baltics Ecosystem Coalition in Tallinn in May 2023 represents a regional commitment to advancing clean technologies (Cleantech Estonia, n.d.). This initiative seeks to accelerate the adoption of clean technologies and promote collaborative efforts across Estonia, Latvia, and Lithuania. By unifying resources and expertise, the coalition aims to transform environmental challenges into business opportunities, enhancing investment potential and shaping policy to nurture the growth of Baltic cleantech startups and scaleups.

These examples illustrate Estonia's multifaceted landscape of sustainable finance from sector-specific innovations to broader regional initiatives, which are vital for forging a path to a sustainable economic future. Yet, improving educational sustainable practices and aligning financial incentives with environmental goals remains a pressing priority for Estonia's sustainable finance journey.

7.5 Recommendations and Summary

Estonia's journey towards integrating sustainable finance practices includes a broad commitment to achieving a resilient and eco conscious future. Although the nation currently lacks a formal state-mandated strategy, the collaborative efforts under the EU-funded TSI project have set the stage for the establishment of "The Knowledge Hub." This proposed structure aims to amplify the understanding and application of ESG criteria across financial and non-financial sectors, addressing a critical knowledge gap that has implications for policy and investment within the country.

The path to a sustainable financial future for Estonia is illuminated by a set of key recommendations aimed at overcoming potential obstacles and capitalizing on transformative opportunities:

1. Sustainable Public Procurement: State and municipal institutions, alongside enterprises, should intensify their focus on sustainable procurement and transition activities within critical sectors such as energy, transport, water management, and data transmission. This approach will set a precedent for integrating sustainability into core operational practices.
2. Expertise and Long-Term Orientation: There is a pressing need for these institutions to deepen their understanding of sustainability. Institutions can prioritize long-term environmental and social outcomes over short-term gains by fostering expertise in this area. Updating procurement practices to value sustainability attributes alongside cost considerations will be essential in this endeavor.

3. Harmonized Sustainability Demands: The current disparate sustainability demands from investors and shareholders can significantly impede unified progress. A harmonized set of expectations, led and enforced by state policy, will alleviate the burden on businesses and streamline the transition to sustainable operations.
4. Principle of Proportionality: Recognizing the diversity of the business landscape, it is crucial to implement the principle of proportionality within sustainability regulations. Expectations should be tailored to the scale and maturity of companies, acknowledging that small and early-stage enterprises may not have the resources to meet the same stringent requirements as their larger counterparts.

Adopting these recommendations will enable Estonia to surmount challenges and harness the full potential of sustainable finance, ensuring that all sectors of the economy contribute to the nation's environmental objectives while fostering equitable economic growth.

References

Cleantech Estonia. (n.d.). https://www.cleantechestonia.ee/

Eesti Pank. (2023). *Estonian Competitiveness Report*. https://haldus.eestipank.ee/sites/default/files/2023-03/kvu2023_eng.pdf

Eesti Pank. (n.d.-a). https://www.eestipank.ee/en/about-us/ngfs

Eesti Pank. (n.d.-b). https://www.eestipank.ee/keskpangast/eesti-panga-strateegia

Estonian Business and Innovation Agency. (n.d.). https://eas.ee/en/

European Commission. (2023a). *Estonia's recovery and resilience plan*. https://commission.europa.eu/business-economy-euro/economic-recovery/recovery-and-resilience-facility/country-pages/estonias-recovery-and-resilience-plan_en

European Commission. (2023b). *Country Report – Estonia*. https://economy-finance.ec.europa.eu/system/files/2023-05/ET_SWD_2023_606_en.pdf

European Investment Bank (EIB). (2023). *EIB Investment Survey: European Union overview*. https://www.eib.org/en/publications/20230285-econ-eibis-2023-eu

Finantsinspektsioon. (2022). https://www.fi.ee/et/investeerimine/aktuaalsed-teemad-investeerimises/finantsinspektsiooni-teekaart-jatkusuutliku-investeerimise-teekonnal-olevatele-ettevotetele

Government of Estonia. (2022). https://valitsus.ee/en/estonia-2035-development-strategy/strategy/strategic-goals

Government of Estonia. (2023, May 11). *Action plan of the Government of the Republic*. https://valitsus.ee/en/media/6506/download

Innovation Center Report. (2022). https://www.fi.ee/sites/default/files/2022-09/Innovatsioon_FI_est_w.pdf

Interreg Baltic Sea Region, https://interreg-baltic.eu/project/sustool/

Kekkonen, A., Pesor, R., & Täks, M. (2023). Stepping towards the Green transition: Challenges and opportunities of Estonian companies. *Sustainability, 15*(5), 4172. https://doi.org/10.3390/su15054172

Kliimamuutused.ee. (n.d.). https://www.kliimamuutused.ee/en

KPMG Baltics, Eversheds Sutherland Bitāns. (2023). *The Sustainable Finance Roadmap for Estonia.* https://www.kik.ee/sites/default/files/sustainable-finance-roadmap-estonia-2023_en_17.08.pdf
LHV. (2023). https://www.lhv.ee/en/green/products
Ministry of the Environment. (2022). *Environmental Awareness Survey of Estonian residents 2022.* MoE, https://envir.ee/kaasamine-keskkonnateadlikkus/keskkonnateadlikkus/uuringud
SmartCap Green Fund. (n.d.). https://smartcap.ee/about-us/sc-gf/.
Staehr, K. (2004). Economic transition in Estonia. Background, reforms, and results. *Contemporary Change in Estonia, 4*, 37–67. https://www.researchgate.net/publication/313607169_Economic_transition_in_Estonia_Background_reforms_and_results
Sustinere. (2022). *ESG and sustainability reporting study.* https://www.sustinere.ee/esg-and-sustainability-reporting-in-the-baltics
Swedbank Estonia. (n.d.). https://www.swedbank.lt/business/finance/loansnewsmallbusiness/farmers?language=ENG
Technical Support Instrument. (2021). https://reform-support.ec.europa.eu/what-we-do/financial-sector-and-access-finance/eu-taxonomy-implementation-and-sustainable-finance-roadmap-estonia-and-latvia_en
The Structure of the Estonian Financial Sector. (2023). https://www.eestipank.ee/en/financial-stability/structure-estonian-financial-sector

Kristiina Kerge is a doctoral candidate at Estonia University of Life Sciences, and consultant in fields of Sustainability, in Estonia.

Chapter 8
Sustainable Finance in the Netherlands

Dirk Schoenmaker and Nathan de Arriba-Sellier

Abbreviations

AFM	Autoriteit Financiële Markten (Dutch Authority for Financial Markets)
AMF	Autorité des Marchés Financiers (French Authority for Financial Markets)
DNB	De Nederlandsche Bank (Dutch Central Bank)
ESG	Environmental, Social, and Governance
SFDR	Sustainable Finance Disclosure Regulation

8.1 Introduction

The financial regulatory framework in the Netherlands follows a 'twin peaks' model, delineating the oversight responsibilities into two distinct domains: conduct of business and prudential regulation (Schoenmaker & Véron, 2021). The Authority for the Financial Markets (Autoriteit Financiële Markten or AFM) oversees the conduct of business, ensuring market integrity and transparency, while prudential oversight, focused on financial solidity and systemic stability, falls under the purview of the Dutch Central Bank (De Nederlandsche Bank or DNB).

Distinct in their approaches to sustainable finance, these two pillars of Dutch financial regulation have carved individual paths. DNB stands out as a precursor of the risks posed by climate change and environmental degradation to the financial system. It has initiated pioneering measures, including implementing the first climate and nature-related stress tests on financial institutions. This leadership role is

D. Schoenmaker (✉) · N. de Arriba-Sellier
Rotterdam School of Management, Erasmus University, Rotterdam, Netherlands
e-mail: schoenmaker@rsm.nl; dearribasellier@rsm.nl

R. Calipha et al. (eds.), *Sustainable Finance Regulation in the European Union*, Sustainable Finance, https://doi.org/10.1007/978-3-031-88392-7_8

partly inspired by the Netherlands' vulnerability to climate-induced phenomena like rising sea levels and flooding. In contrast, the AFM, while exercising caution, has been progressively assertive in response to emerging EU regulations regarding transparency and business conduct. However, the AFM has refrained from aggressively leveraging its authority in sustainable finance. This is particularly noteworthy given the Netherlands' status as a financial nexus, boasting the largest pension market in Europe and ranking as the continent's fourth largest financial center surpassed only by the United Kingdom, France, and Germany. The significance of the Dutch financial sector was underscored in 2021 when Amsterdam momentarily ascended to become Europe's premier share trading hub. Against the backdrop of the Netherlands' strategic financial position, the interplay between these two regulatory bodies sets a complex stage for the evolution and enforcement of sustainable finance within a key EU member state. With financial regulation mainly being the product of European Union legislation since the Global Financial Crisis, national authorities are mostly left with supervisory roles and little discretion. Still, the AFM and the DNB have wisely used their responsibilities and paved the way in many respects for more assertive European financial supervision about sustainable finance.

This chapter focuses on the distinctive approaches of the Dutch supervisory authorities towards sustainable finance. It begins by examining the AFM's initial conservative stance (Sect. 8.2), proceeds to explore how the authority has started to assert its regulatory philosophy (Sect. 8.3), and then discusses its gradual shift towards a more enforcement-driven approach (Sect. 8.4). Throughout this chapter, the DNB's regulatory efforts warrant special attention. The DNB's strategies are illustrative of its commitment to mitigating environmental risks and reflect the spectrum of approaches that supervisory authorities can adopt in addressing the challenges of sustainable finance. Through this comparative analysis, the last section aims to shed light on the evolving dynamics of financial supervision in the Netherlands and its implications for the future of sustainable finance regulation in the European Union (Sect. 8.5) (European Commission, EC, 2018).

8.2 Learning by Watching: The AFM's Cautious Approach to Sustainable Finance

As sustainable finance gained steam from increasingly interested investors, the AFM sought first to observe the phenomenon rather than seeking to support or regulate it. Still, the AFM did not start considering sustainable finance until after the EU published the final report of the High-Level Expert Group on Sustainable Finance, which opened a more aggressive path of promoting sustainable finance through regulation (EC, 2018).

Surveying the Market

In 2020, the AFM published its report on sustainable bonds in the Netherlands. It reported a surge in the issuance of sustainable bonds from EUR 9 billion in 2018 to EUR 18 billion in 2019 (AFM, 2020a). As a result, the Netherlands was the fifth-largest market of sustainable bonds worldwide. The main conclusion drawn by the AFM from its in-depth study was that there was a significant risk for greenwashing as transparency and standardization were lacking in the sustainable bond market. These risks led the AFM to commit to increasing its scrutiny, under the EU Prospectus Regulation, over the provision of sustainability-related information by issuers and the advertisement of sustainable bonds (EC, 2023). It even promised enforcement action if the information provided by issuers proves incorrect, unclear, or misleading (EC, 2023). Still, concerning regulation, AFM merely emphasized its endorsement of standardization as undertaken by the EU Green Bonds Standard Regulation and its intention to "continue to monitor regulatory developments" (AFM, 2020a, 2020b, 2021, 2022a; Arriba-Sellier & Van Caenegem, 2025; EC, 2023). No specific regulatory action, or even guidance, was announced. More recently, in November 2022, the AFM released the results of an inquiry into retail investors' interest in sustainability (AFM, 2022a). The comprehensive study sought to map out the motives of sustainability-oriented investors, their choices, the retail investment process they go through, and the obstacles they face. The results show that "impact" was the most critical investment objective for nearly half of those retail investors. At the same time, ethics was cited by nearly a third, and returns came third with slightly over 20% of the investors surveyed (AFM, 2020b). The study showed contradictions in investment choices as chosen investment strategies and investments fall short of the desired impact investments (AFM, 2020b). Lack of (reliable) information is pointed out by the AFM as an obstacle for these retail investors, which lists among the factors leading to investors' confusion as limited financial literacy as much as distrust in financial intermediaries, the lack of suitable products, and misleading marketing claims (AFM, 2020b). Still, the study's conclusions are silent about the lessons or potential responses the AFM could draw for regulation and supervision.

Market Guidance in the Implementation of SFDR

The legislative initiatives at the EU level, particularly the Taxonomy Regulation, the Sustainable Finance Disclosure Regulation (SFDR), and the amendments to the Benchmarks Regulation, served as a pivotal moment for the Authority for the Financial Markets (AFM). These developments necessitated a departure from the AFM's erstwhile restrained engagement with sustainable finance, ushering in an era of active regulatory participation. Responding to these legislative stimuli, the AFM issued a seminal 'dear CEO letter' in June 2020 directed at asset managers. This correspondence aimed to raise awareness about the imminent changes in the regulatory landscape and to outline the expectations for compliance. While this initial communication did not prescribe supervision or enforcement actions, it provided crucial market guidance, signaling a shift in the AFM's approach from passive oversight to active facilitation (AFM, 2020b). The timing of this letter, aligned with the adoption of the Taxonomy Regulation, underscores the strategic intent to emphasize the SFDR—legislation that had been established 6 months prior. The AFM's concentrated focus on the SFDR indicates the complexity and implementation challenges associated with the regulation, which have since become more pronounced (AFM, 2020b).

The letter reflects the broader regulatory and supervisory uncertainties that have emerged recently (Arriba-Sellier & Van Caenegem, 2025). It highlights the intricacies of interpreting SFDR Articles 8 and 9, which asset managers misconstrued as labeling requirements rather than the intended enhancement of disclosure practices. The AFM's letter serves as a barometer for the evolving expectations of sustainable finance disclosures within the EU's financial markets by addressing potential misconceptions around marketing and product focus. The AFM's letter was followed by a second one, addressed to pension funds, which were initially overlooked. Pension products are indeed financial products that fall within the scope of SFDR. This second letter follows the same logic, i.e., to provide market guidance in advance of the implementation of SFDR, clarify certain issues related to the obligations of pension funds as well as respond to the industry's questions over compliance. In particular, the letter draws the attention of the addressees to the consequences of their (voluntary) commitments under the Covenants for International Corporate Social Responsibility (IMVO) and Socially Responsible Investments (IMVB), which imply that they consider negative impacts on sustainability in their investments and hence need to provide sufficient information in this respect. As SFDR started applying in March 2021, AFM changed its approach. It started surveying the market's compliance and issuing recommendations to this end. The first AFM study of SFDR's implementation was released in September 2021 and focused on asset managers (AFM, 2021). In the absence of technical standards and regulatory certainty over the interpretation of certain requirements, AFM clarified that this study aimed at monitoring compliance but did not amount to stricter supervision or even enforcement. It surveyed 1250 funds, 57% of which have neither sustainability characteristics nor a sustainability objective (under Articles 8 and 9 of SFDR, respectively), 35% promote sustainability characteristics, and 8% have a sustainability objective. According to AFM, all managers of funds that consider sustainability in one way or another provided related information in pre-contractual disclosures. Still, the AFM criticized both the "classification" of some funds and the quality of the information provided by fund managers. In particular, AFM warned against the overly general information provided by asset managers on sustainability risks to which their funds are exposed. This obligation applies to all funds, regardless of their characteristics, following Article 6 SFDR. In addition, AFM criticized fund managers that withhold clarity on their integration of sustainability considerations and, for financial products that fall under Article 9 SFDR, the funds' objectives.

The AFM followed up in November 2022 with a second series of studies, distinguishing between asset managers, the pension sector, and other market participants. The AFM also brought compliance with the Taxonomy within the scope of its study as it became relevant for compliance with SFDR after the Taxonomy started to apply (AFM, 2022b). The AFM's study shows significant expansion of the disclosure provided by asset managers, investment firms, banks, and insurance companies, even though it continues to criticize the general character of the information disclosed and the lack of concrete insight it gives, such as remuneration policies. Moreover, it exposes significant changes in the "classification" of funds with fewer funds that have sustainability as an objective, although practically no change for funds considered to fall under Article 8 SFDR. Regarding the EU Taxonomy, the AFM surprisingly reported that none of the funds integrating sustainability considerations, except for one, invest in environmentally sustainable activities as defined by the Taxonomy Regulation. However, the lack of reliable data is cited as a factor explaining this absence. The conclusion of the AFM is similar to the first study despite the clarified legal situation. As such, it did not announce particular measures to provide stricter supervision or enforcement of the SFDR.

8.3 The Limited Regulatory Role of AFM

Despite its reserved stance on taking a direct regulatory role in sustainable finance or advocating for the Dutch Ministry of Finance to enact pertinent legislation, the Authority for the Financial Markets (AFM) has actively shaped EU policy and legislative frameworks. Collaborating with its French counterpart, the Autorité des Marchés Financiers (AMF), it contributed to the European discourse on sustainable finance through the issuance of position papers concerning the regulation of sustainable bonds and ESG (Environmental, Social, and Governance) ratings ahead of the European Commission's legislative proposals in these domains. To refine the discourse further, the AFM recently unveiled its position paper aimed at enhancing the Sustainable Finance Disclosure Regulation (SFDR), following the contentious recommendations by its French counterpart.

AFM as a Stakeholder

In 2018, in anticipation of the European Commission's Action Plan on sustainable finance, the AFM and the French AMF co-authored a position paper to influence the regulatory approach towards green, social, and sustainable bonds (Autoriteit Financiële Markten (AFM) & Autorité des marchés financiers (AMF), 2019). This document offered early insights into the authorities' stance on the need for regulation in sustainable finance. This view not only predated the EU Green Bonds Standard proposal but also the AFM's analysis of the Netherlands' sustainable bonds market. Contrary to advocating for a specific prospectus for green bond issuers, the position paper acknowledged the detrimental effect of the lack of mandatory EU-level regulation on green bonds to market expansion. It proposed a solution whereby issuers tasked with classifying their bonds as green, social, or sustainable, would also be mandated to disclose specific information regarding the usage of bond proceeds. This included adherence to standards, intentions to report on proceed utilization, and the provision of third-party verification.

Notably, the supervisory bodies' recommendations exceeded the initial and final iterations of the EU Green Bonds Standard Regulation by suggesting more stringent requirements than the EU legislature's preference for a voluntary, though rigorous, framework for green bond issuers. This discrepancy underscores a critical gap identified by the AFM (and echoed by its own 2020 report) and the AMF—the insufficient information available to investors concerning the utilization of bond proceeds, potentially misleading them. The AFM illustrates a nuanced approach to its regulatory role in sustainable finance through these initiatives. While not directly imposing regulations, it endeavors to influence broader EU policies and standards to bridge informational gaps and foster a more transparent, reliable, and sustainable finance market.

Persisting in their efforts to shape sustainable finance regulation, the Authority for the Financial Markets (AFM) and its French counterpart, the Autorité des Marchés Financiers, proactively issued a position paper in late 2020. This document made a compelling case for the European Union to regulate the provision of ESG (Environmental, Social, and Governance) data, ratings, and related services AMF & AFM, 2020. Notably, this advocacy came ahead of the European Commission's announcement of a renewed sustainable finance strategy, which included a proposal for regulating ESG ratings (EC, 2021). Unlike the earlier focus on EU Green Bonds, this position paper embraced a wider ambit, calling for a comprehensive regulatory framework encompassing not only ESG rating agencies but also all entities involved in providing sustainability-related data services.

The AFM and AMF underscored the importance of applying regulatory measures across the board to ensure transparency, especially concerning the methodologies used by ESG rating and data service providers. They argued that limiting regulation solely to ESG ratings could lead to regulatory arbitrage, undermining the effectiveness of the proposed framework. Despite their comprehensive approach, the European Union's legislature ultimately opted for a more focused regulatory strategy, concentrating on ESG ratings alone.

However, the legislative proposal put forth by the European Commission aligns with several key recommendations from the AFM and AMF (EC, 2023). It mandates transparency in the methodologies of ESG ratings, sets forth stringent requirements on governance and internal controls, and introduces safeguards against conflicts of interest. Additionally, it assigns the European Securities Markets Authority (ESMA) the responsibility to oversee ESG rating agencies, paralleling ESMA's existing supervisory role over credit rating agencies. This alignment indicates a significant step towards integrating AFM and AMF's insights into the evolving EU regulatory landscape for sustainable finance despite differences in the scope of regulation.

AFM's Guidelines on Green Claims

The AFM's historically cautious approach to sustainable finance, predominantly characterized by its supervisory role within the framework of EU financial regulations, saw a pivotal shift in 2023. Marking its most significant and direct regulatory action in the sphere of sustainable finance to date, the AFM proactively issued guidelines on green claims (AFM, 2023a). This intervention stands out as a notable example of regulatory action explicitly addressing sustainability claims within the financial sector.

The guidelines issued by the AFM target financial institutions and pension providers, offering directives on how to present sustainability claims accurately. These guidelines were introduced in response to concerns over greenwashing observed in the implementation of the SFDR. Yet, they do not confine themselves to a specific legislative context (such as investment services). Instead, their scope is deliberately broad, resonating with the EU's legislative endeavors concerning environmental claims and consumer protection.

Defined broadly, sustainability claims encompass any form of expression market participants use to describe and promote their entities or products and services, including labels, images, certificates, targets, statements, etc. The AFM does not narrow the definition of sustainability; rather, it encompasses a wide array of issues touching upon environmental, social, or governance (ESG) factors, extending to human rights, ethics, and stakeholder relations. Notably, the AFM clarifies that its conception of sustainability extends beyond the definitions outlined in the SFDR, MiFID II, or the Taxonomy Regulation, aligning more closely with the expansive view introduced by the Corporate Sustainability Reporting Directive (CSRD).

Operating on principle-based guidelines, the AFM's directives are context-sensitive, aiming to guide the financial industry in providing accurate information to consumers and ensuring compliance with legal obligations. Although these guidelines constitute an act of soft law—lacking the binding legal force of formal laws and regulations—they play a critical role in aiding financial institutions and pension providers to align with their disclosure and marketing duties under both EU legislation (including SFDR, MiFID II, and the Regulation on the Cross-Border Distribution of Funds) and Dutch national laws (such as the Financial Services Act, the Unfair Commercial Practices Act, and the Pensions Act).

Through these guidelines, the AFM seeks to elevate the standard of transparency and integrity in the marketing and disclosure of sustainability claims, thereby supporting the financial sector's compliance efforts and contributing to the broader objective of combating greenwashing.

Principle 1: Accuracy and Currency of Sustainability Claims

Principle 1 underscores the imperative for sustainability claims to be accurate, representative, and current. It mandates that market participants ensure their sustainability assertions are factually correct and consistent with other disclosed information. An example cited by the AFM involves a contradiction where a firm claims to overlook the adverse sustainability impacts of its investments while simultaneously supporting the Covenant on International Corporate Social Responsibility, illustrating a misuse of sustainability claims. Furthermore, claims must provide a truthful representation of the entity's or product's sustainability attributes. For instance, marketing that overstates a commitment to sustainability or imagery suggesting environmental stewardship while engaging in environmentally harmful practices, such as investing in companies reliant on deforestation, contradicts this principle. Lastly, the principle demands that claims remain current, cautioning against reliance on obsolete data, like outdated ESG ratings, to support sustainability assertions. This principle seeks to ensure that sustainability claims reflect the entity's practices and policies, fostering transparency and preventing misinformation.

Principle 2: Clarity and Evidence in Sustainability Claims
Principle 2 demands that sustainability claims be explicit and well-supported. Claims must detail their specific impact on the product or entity, avoiding vague statements that lack quantifiable information or clear descriptions of sustainable investment policies. Furthermore, these claims must be backed by solid, verifiable evidence. The AFM highlights the necessity for transparency, insisting that unsubstantiated claims be avoided. For Example, broad statements about net-zero or biodiversity goals should include concrete baseline data, strategic pathways, and implementation details. This principle ensures that sustainability assertions are precise and rooted in credible and current information, fostering trust and accountability in sustainability practices.

Principle 3: Clarity, Relevance, and Accessibility of Sustainability Claims
This principle requires that information is presented in clear, simple language, elucidating any complex concepts or legal jargon to ensure comprehensibility. An example provided by the AFM criticizes overly complex explanations that obscure the investment policy's essence. Additionally, terminology in sustainability claims must be precise and align with stakeholders' expectations, such as reserving the term "impact" exclusively for investments with a measurable positive effect on sustainability. Moreover, all pertinent information supporting a claim should be straightforward to locate, avoiding the dispersal of critical data across lengthy and intricate documents. This principle aims to enhance the transparency and efficacy of sustainability communications, ensuring they are both informative and user-friendly.

The AFM aims to foster a more transparent and trustworthy sustainable finance environment through these principles. It encourages market participants to present their sustainability efforts honestly and accurately to prevent green-washing.

8.4 Enhanced Supervision and Enforcement in Sustainable Finance

After adopting a cautious stance on sustainable finance for several years, the Authority for the Financial Markets (AFM) is shifting towards a more assertive supervision and enforcement role, spurred by the European Union's legislative developments. The recent period, previously focused on offering market guidance and clarifying the Sustainable Finance Disclosure Regulation (SFDR), has evolved. With the SFDR now in effect and its technical standards operational, asset managers are compelled to adjust. The AFM allowed a 6-month grace period for financial product providers to meet the latest technical standards, after which it began actively supervising compliance.

Adapting New Norms: The AFM and Sustainable Reporting Obligations

The AFM's guidelines on sustainability claims underscore the authority's growing impatience and increased vigilance over compliance within the asset management and pension sectors. Additionally, the conclusion of the Taxonomy Regulation's introduction phase and the activation of delegated acts defining the technical screening criteria for the EU Taxonomy compel companies and financial institutions to adhere to their disclosure responsibilities. Prompted by an ESMA study revealing that 40% of surveyed companies had incomplete or missing taxonomy-related information, the AFM has intensified its call for supervised entities to address this issue, signaling a pivotal turn towards enforcement and enhanced risk management in the realm of sustainable finance.

As the Corporate Sustainability Reporting Directive (CSRD) takes effect and the European Sustainability Reporting Standards (ESRS) come into play, the AFM emphasized the importance of adhering to the evolving reporting requirements. The authority has cautioned companies about the urgency of gathering sustainability data promptly to meet the CSRD mandates. It has also highlighted businesses' need to transparently report their environmental and societal impacts, adhering to the CSRD's double materiality principle. Following an evaluation of climate reporting by 27 companies, the AFM identified a prevalent opacity in disclosing impacts. For the 2023 reporting cycle, it urged firms to clarify their net-zero ambitions, including detailed transition strategies, scope definitions, the inclusion of greenhouse gases beyond CO2, and the application of carbon credits. However, the AFM's stance on prioritizing the enforcement of these obligations remains to be fully articulated despite the significant implications of the CSRD's application.

Enhanced Scrutiny on Environmental Risks

The AFM initially recognized climate-related risks in its early sustainable finance publications in 2020. However, only recently has it significantly incorporated this focus into its operational agenda. In 2024, the AFM's risk mapping highlighted various sustainability challenges, notably emphasizing the real estate sector's vulnerability due to inadequate pricing of climate risk exposures. A pivotal study released by the AFM in 2023 shed light on the substantial financial jeopardy buyers and homeowners face from climate-induced damages, with uninsured climate related repair costs averaging €54,000 (AFM, 2023b). The authority advocated for greater transparency and increased reserve requirements for purchasers of at-risk properties. However, this concern for mispricing and associated risks has not been extensively applied to other economic sectors or commercial real estate, with the AFM's risk analysis primarily spotlighting the residential housing market.

Additionally, the AFM identified a medium risk in capital markets stemming from a lack of sustainability information while categorizing greenwashing as a significant concern across asset management and accounting sectors. Following regulatory updates to the UCITS and AIFM directives concerning investment funds, the AFM disclosed findings from a survey on asset managers' sustainability risk management practices. This survey revealed difficulties distinguishing between sustainability objectives and risks, along with a varied commitment to due diligence for identifying sustainability risks (AFM, 2023c).

8.5 Concluding Remarks: Comparison with the DNB

This chapter underscored the divergent paths in the regulatory and supervisory approaches to sustainable finance in the Netherlands, highlighting the dynamic interplay between transparency requirements and prudential oversight in navigating the challenges and opportunities presented by environmental sustainability in the financial sector.

Transparency vs. Prudential Regulation in Sustainable Finance Initially, the domain of transparency in sustainable finance embarked on a regulatory path that has progressively embraced supervision and enforcement. Conversely, the prudential sphere has charted a different course. The Basel Committee on Banking Supervision, the cornerstone of prudential regulation, has been slow to adapt, starting with the issuance of principles for managing and supervising climate-related financial risks before contemplating regulatory adjustments to address these risks (Gruenewald et al., 2023). This cautious approach contrasts sharply with the proactive stances of prudential supervisors like the European Central Bank (ECB), the Bank of England, and the Banque de France, along with the Dutch Central Bank (DNB), which have positioned themselves as leaders in incorporating climate risks into their supervisory frameworks. In 2022, the ECB integrated climate risk evaluations into its Supervisory Review and Evaluation Process (SREP) under the Basel Capital Adequacy Framework's Pillar 2, opting for qualitative feedback and measures without imposing additional capital requirements. The anticipation of integrating heightened capital charges for environmental risks under Pillar 1 highlights a critical yet unmet step toward strengthening the regulatory framework.

Comparing AFM and DNB The AFM has steadily enhanced its sustainability focus. Yet, it is the Dutch Central Bank (DNB) that has emerged as a trailblazer in the realm of sustainability supervision in collaboration with the Bank of England and the Bank of France (Banque de France). As inaugural members of the Network for Greening the Financial System (NGFS) in 2017, these institutions have spearheaded efforts to integrate sustainability considerations into financial supervision. Today, this network include 134 members.[1] DNB, in particular, has demonstrated commendable initiative by conducting early research on climate, water, and biodiversity risks and by orchestrating climate stress tests on the Dutch banking sector.[2] Furthermore, DNB's establishment of a national platform for sustainable finance in 2016 bringing together financial institutions, regulatory bodies, government entities, and academia has been pivotal in fostering sustainable finance initiatives across the Netherlands.

References

AMF & AFM (2020). Position Paper: Call for a European Regulation for the provision of ESG data, ratings, and related services, https://www.amf-france.org/sites/institutionnel/files/private/2020-12/amf-afm-position-paper-call-for-a-europeanregulation-for-providers-of-esg-data-ratings-and-related-services.pdf

AFM (2023a). Guidelines on Sustainability Claims, https://www.afm.nl/en/sector/actueel/2023/oktober/leidraadduurzaamheidsclaims

AFM (2023b). Inprijzen klimaatrisico's op de Woningmarkt, https://www.afm.nl/nlnl/consumenten/actueel/2023/november/inprijzen-klimaatrisicoschade

AFM (2023c). Duurzaamheidsrisicobeheersing, https://www.afm.nl/nl-nl/sector/actueel/2023/augustus/verkenning-duurzaamheid-risico-management

Arriba-Sellier, N., & Van Caenegem, A. (2025). Sustainability-related materiality in the SFDR. In K. Alexander, M. Siri, & M. Gargantini (Eds.), *The Cambridge handbook of European sustainable finance*. Cambridge University Press.

Autoriteit Financiële Markten (The Dutch Authority for the Financial Markets). (2020a, April). *Report on sustainable bonds in the Netherlands: Developments, risks and impact on supervision.*

Autoriteit Financiële Markten (The Dutch Authority for the Financial Markets). (2020b, June). *Sectorbrief over duurzaamheid.*

[1] NGFS website accessed 8 January 2024: https://www.ngfs.net/en/about-us/membership.

[2] Platform voor duurzame financiering, at https://www.dnb.nl/en/green-economy/sustainable-finance-platform.

Autoriteit Financiële Markten (The Dutch Authority for the Financial Markets). (2021, September). *Implementatie SFDR: Onderzoek naar implementatie van SFDR vereisten door beheerders van Nederlandse beleggingsinstellingen.*

Autoriteit Financiële Markten (The Dutch Authority for the Financial Markets). (2022a, November). *Duurzame beleggers in kaart: Onderzoek naar doelstellingen en verwachtingen van duurzame retailbeleggers.*

Autoriteit Financiële Markten (The Dutch Authority for the Financial Markets). (2022b, November). *SFDR in de praktijk: blijvende aandacht nodig. Eerste onderzoek naar naleving van SFDRen Taxonomie Vereisten door pensioenfondsen en PPI's.*

Autoriteit Financiële Markten (The Dutch Authority for the Financial Markets) & Autorité des marchés financiers (The French Authority for the Financial Markets). (2019, April). *Position paper on green, social, sustainable bonds.* https://www.afm.nl/en/sector/actueel/2019/apr/transparantie-prospectus-groene-obligaties

EC (2021). Strategy for Financing the Transition to a Sustainable Economy, COM(2021)390 final, https://eurlex.europa.eu/legal-content/EN/TXT/?uri=celex:52021DC0390

European Commission. (2023). Proposal for a regulation of the European Parliament and of the Council on the transparency and integrity of Environmental, Social and Governance (ESG) rating activities, COM (2023) 314 final.

European Commission, High-Level Expert Group on Sustainable Finance. (2018). *Financing a European economy, final Report.* https://finance.ec.europa.eu/system/files/2018-01/180131-sustainable-finance-final-report_en.pdf

Gruenewald, S., Knijp, G., Schoenmaker, D., & Tilburg, R. (2023). Embracing the brave new world: A response to Demekas and Grippa. *Journal of Financial Regulation.* https://doi.org/10.1093/jfr/fjad011

Schoenmaker, D., & Véron, N. (2021). A twin peaks vision for Europe. In A. Godwin & A. Schmulow (Eds.), *The Cambridge handbook of twin peaks financial regulation* (pp. 282–291). Cambridge University Press.

Dirk Schoenmaker is Professor of Banking and Finance at the Rotterdam School of Management, Erasmus University Rotterdam Netherlands, focusing on sustainable finance, central banking, financial stability, financial system architecture, and European financial integration. Schoenmaker is also a Research Fellow at the Centre for European Policy Research (CEPR).

Nathan de Arriba-Sellier is a researcher and the Director of the Erasmus Platform for Sustainable Value Creation at Rotterdam School of Management, Erasmus University, Netherlands, focusing on sustainable finance, law and policy.

Chapter 9
Israel: Local Trends in Sustainable Finance

Dafna M. DiSegni, Nadav Paldi, and Gal Tamir

Abbreviations

ANSI	Adjusted Net Saving index
DNSH	Do No Significant Harm
ESG	Environment Society and Governance
GGI	Green Government's Initiative
ICMA	International Capital Market Association
ICPA	Israeli Institute of Certified Public Accountants
IEC	Israel Electric Corporation
ISA	Israeli Security Authority
MoEP	Israeli Ministry of Environmental Protection
NGFS	Network of Central Banks for Greening the Financial System
TCFD	Task Force on Climate-related Financial Disclosure
TEG	Technical Expert Group

D. M. DiSegni (✉)
University of Rhode Island, Kingston, RI, USA
e-mail: dafna.disegni@uri.edu

N. Paldi
Ministry of the Environment, Jerusalem, Israel

G. Tamir
Israel National Digital Agency, Strategy and Policy Unit, Tel Aviv, Israel
e-mail: galt@digital.gov.il

R. Calipha et al. (eds.), *Sustainable Finance Regulation in the European Union*, Sustainable Finance, https://doi.org/10.1007/978-3-031-88392-7_9

9.1 Introduction

Israel's distinctive journey towards a sustainable financial path is marked by necessities, lack of natural resources, and communal enterprise of a building-up society in Israel. Israel's population is compact yet diverse, with a size that surpasses 9 million residents. This population is a tapestry of cultural and ethnic, including Jews, Arabs, Druze, Bedouins, and other groups, each contributing to the rich social fabric of the nation (Vinrab & Shraverman, 2023). Despite its relatively small geographic size (21,000 sq. km), Israel's demographic composition and lack of natural resources have made it a hub of innovation and economic activity (Lotan et al., 2018).

The sustainable finance architecture of Israel is influenced by its history of communal living epitomized by the 'kibbutzim' (collectives) established in Israel since the beginning of the twentieth century (Curtis & Chertoff, 1973; Sibylle, 2010). These collective communities have not only shaped social interactions but have also contributed to a culture of shared sustainability goals (primarily UN Sustainability Goals 1, 2, 5, 8, 16, 17), and the development of agricultural technologies (from irrigation and soil cultivation to seed variety and quality), water industries, and new economic sectors, designed to integrate communities into the labor force, despite the lack of resources. In this context of a nation in development since 1948 till today, sustainable finance is not just an approach but a necessity (Fig. 9.1). Data is presented as a percentage of gross national income. We present the World Bank Adjustment Saving Index of Sustainability on the horizontal axis. On the vertical axis, we present the breakdown of the Index computation by components for each specified year. Enjoy society, and a culture. By exploring the intersection of development, finance, and sustainability, this chapter will highlight the initiatives,

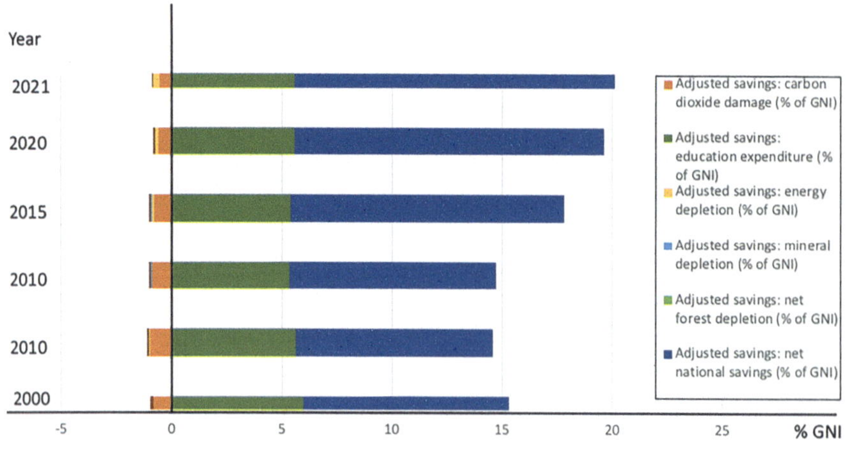

Fig. 9.1 Israel Adjusted Net Saving Index for 2000–2021, in percents of gross national income. (Data Source: World Bank database 2023). On the horizontal axis: World Bank Adjustment Saving Index of Sustainability in % of GNI. On the vertical axis: the Index components for each specified year

policies, and financial frameworks that have enabled Israel's success in transforming its limitations into opportunities, securing its place as a sustainable growth and resilience model.

A glance at Israel's Adjusted Saving Index of sustainability suggests that the state's sustainability is primarily based on its Produced Capital (Net National Savings) and Human Capital over two decades. Figure 9.1 details the components of the World Adjusted Net Saving Index (ANSI), including Produced Capital (measured by Net National Savings), Human Capital (measured by Investment in Education), and Natural Resources Capital (Including values of changes in Carbon emissions, forest reduction, minerals reduction, energy resources depletion, and particulate emissions (see Bolt et al., 2002 for explanation on the World Bank Adjusted Saving Index calculation). Moreover, we observe that in Israel, natural resources are not an engine of growth nor of significant impact on the nation's sustainability, at least not until the year 2016 when a significantly large reserve of Natural Gas was found off-shore in the Mediterranean Sea, within the economic zone of Israel (the Israeli Leviathan natural gas reserve).

The statutory entities that supervise de facto the financial system in Israel and determine the local regulation are the Bank of Israel, which oversees the banks; Capital Market, Insurance and Savings Authority that supervises non-bank financial entities insurance companies, pension funds, provident and training funds, and non-bank credit (In Israel, the saving pension is mandatory); The Securities Authority oversees the stock exchange and public companies. These together shaped the sustainable finance system observed today.

9.2 Development of Israel's Sustainable Finance Strategy

Over the past two decades, Israel has witnessed a remarkable evolution in the landscape of Sustainable Finance considerations, reflecting a growing awareness of the critical interplay between financial markets and environmental, social, and governance (ESG) factors. In 2005, the Tel Aviv Stock Exchange introduced the Maala ESG index, a pioneering move that laid the foundation for integrating sustainability into investment criteria.[1] Subsequently, regulatory bodies took noteworthy steps to address environmental risks, with the Capital Market, Insurance and Saving Authority, and the Banking Supervision Department of the Bank of Israel issuing general instructions for risk management related to environmental law violations in 2007 and 2009, respectively. While these early guidelines focused primarily on risk mitigation, they set the stage for more comprehensive ESG integration.

[1] For more information on MAALA index please see: https://market.tase.co.il/en/market_data/index/150/graph.

Disclosure

The year 2011 marked a significant milestone as the Ministry of Environmental Protection, in collaboration with financial regulators, published a guide outlining environmental risk management strategies in investment, emphasizing the importance of aligning financial decisions with sustainability goals. The Banking Supervision Department mandated the Banks to publish corporate responsibility policy reports every two years, heralding a broader commitment to corporate social responsibility. In 2012, the Israel Securities Authority introduced legislation compelling companies to disclose material environmental risks in their prospectuses, underscoring the importance of transparency in sustainability. By 2014, the Ministry of Environmental Protection had introduced the Environmental Impact Index, which annually ranks public companies based on their environmental impact and adherence to environmental laws, highlighting the increasing significance of sustainability performance metrics.

The subsequent years witnessed a more concerted effort toward ESG integration. In 2017, the Capital Market, Insurance, and the Saving Authority issued adoption and disclosure instructions for institutional investment firms, cementing ESG as an integral part of the financial landscape. The year 2020 marked Israel's global commitment to green finance, with the Bank of Israel joining the Network of Central Banks for Greening the Financial System, aligning its efforts with the broader international sustainability movement. This momentum continued into 2021, with the Tel Aviv Stock Exchange joining the Sustainable Stock Exchanges Initiative and establishing the "Forum +35 for Gender Diversity" by the Israel Securities Authority, promoting greater gender diversity on corporate boards.

The 'Israeli Green Finance Forum' was established in 2020 to enable knowledge sharing between the various state actors involved in Green Finance, whether in regulatory and stability aspects or aspects of complementary policy tools such as the formation of knowledge and training. The forum includes members of the three Israeli financial authorities: the Capital Market, Insurance and Savings Authority, the Securities Authority, and the Central Bank of Israel (Banking Supervision Department), as professionals from the financial sectors, as well as representatives from various government ministries, including the Ministry of Finance, The Ministry of Justice, the National Economic Council and the Ministry of Environmental Protection. The year 2022 represented a turning point, as Israel embarked on initiatives such as developing its Taxonomy, closely modeled on the EU Taxonomy, and the publication of a National Climate Physical Risks Map and Web Portal. Additionally, regulatory provisions for ESG disclosures in banks and mutual funds were introduced, reflecting a commitment to comprehensive ESG integration across the financial sector (adopting Task Force on Climate-related Financial Disclosure (TCFD)) and this period also witnessed the Ministry of Finance's groundbreaking Green Bond Framework, paving the way for sovereign green bond issuances in the international market.

The Israeli Taxonomy as a Tool of SF Accelerator

The Israeli Mitigation Taxonomy, launched in 2024 is a systematic tool for classifying economic activities according to their influence on climate change mitigation and other environmental goals, and it follows the EU Taxonomy.[2] The fundamental objectives of the Israeli mitigation Taxonomy are to provide a transparent framework for identifying economic activities that make substantial contributions to mitigation, to establish standardized definitions for sustainable investments, and to prevent green-washing. In other words, based on the mitigation Taxonomy, financial players such as investors, companies, and issuers can define and identify various activities as significantly contributing to climate change mitigation ("activities aligned with the Taxonomy") using a standardized professional language.

The Taxonomy is designed to guide the market towards sustainability and encourage environmental innovation and enhancement. Consequently, the Taxonomy establishes demanding criteria for a substantial environmental contribution. Therefore, economic activities must surpass standard greenhouse gas emissions reduction measures to be acknowledged as aligned with the mitigation Taxonomy. A lack of alignment with the Taxonomy does not automatically imply that economic activity is significantly harmful to the environment; instead, it simply does not meet the standards of outstanding environmental performance for financial activities.

The methodology behind the Israeli Mitigation Taxonomy follows the European Taxonomy (European Commission, 2020b). The European Taxonomy is rooted in European Union Parliament legislation and delegated acts established by the European Commission. According to the European Taxonomy regulation (European Commission, 2020b), financial entities and select non-financial companies within the European market are mandated to publish reports detailing the extent of compliance of their investments and economic activities with the established Taxonomy. Moreover, the EU introduced disclosure obligations for financial and non-financial companies, a directive that came into effect in January 2022.

In contrast to the European Taxonomy, the Israeli Mitigation Taxonomy presently operates as a voluntary standard. Consequently, financial entities, investment organizations, and real companies are not compelled to report their alignment to the Taxonomy. This is a fundamental disadvantage of the Israeli mitigation system since without incentives on the one hand for promoting compliance and sanctions against non-compliers, it will be a lot more difficult to make those actors participate in the Taxonomy system. Therefore, a significant aspect of the plans will inevitably involve establishing an incentive system and regulatory framework to encourage engagement with the Taxonomy.

The European Taxonomy Regulation from 2020 defines six environmental objectives (European Union Commission Delegated Regulation, 2020a, 2020b): (1)

[2] For more information about the EU taxonomy: https://finance.ec.europa.eu/sustainable-finance/tools-and-standards/eu-taxonomy-sustainable-activities_en.

mitigation; (2) adaptation; (3) sustainable use of water and marine resources; (4) transition to a circular economy; (5) pollution prevention and control; (6) protecting biological diversity and ecosystems. While in Europe, all six parts of the Taxonomy have already come into effect, in Israel, only the mitigation Taxonomy has been officially launched, and the adaptation Taxonomy is anticipated to be released soon for public consultation. Subsequently, there is an expectation to adopt all six taxonomies in Israel. The development of the Mitigation Taxonomy was initiated at the end of 2021 and was led by the Israeli Ministry of Environmental Protection (MoEP). After thorough discussions with diverse international entities, it became evident that the European Union is at the forefront of financial sustainability, especially with its European Taxonomy. Consequently, the choice was made to anchor the Israeli Taxonomy in the methodology of the European Taxonomy, with essential modifications tailored to Israel and its market. The initial phase of formulating the Israeli mitigation Taxonomy involved consultations with a team of experts from the European Union, who were part of the Technical Expert Group responsible for drafting the European Taxonomy.[3] This stage also included discussions with the Israeli Green Finance Regulators Forum.[4] In the subsequent phase, a steering team, including representatives from financial regulators and government ministries in Israel, played a crucial role in refining the methodology of the Taxonomy and the required adjustments. Additionally, extensive consultations were conducted with various professional divisions within the Ministry of Environmental Protection. The process was led by EcoTraders, an Israeli consulting company specializing in economic and environmental subjects. In October 2022, the Taxonomy was released for public and stakeholder consultation.[5] It received extensive feedback from academia, the private sector, and government entities throughout this phase. Notably, most of these comments led to revisions and enhancements in the final document. The Israeli Taxonomy is directed to the following economic sectors: Industry and Trade, Energy, Water and Waste, Transportation, Construction, Communication, and Education (professional, scientific, and technical activities). Within these sectors, there are 74 economic activities. To be aligned with the Taxonomy, these sectors must meet three cumulative conditions (as similarly requested by the EU Taxonomy):

[3] Technical guidelines and policy by the European Univen Financial Stability, Financial Services and Capital Markets Union see: https://commission.europa.eu/document/download/b5a1053c-396d-416f-83c8-db1a3d5d597f_en?filename=fisma_sp_2020_2024_en.pdf.

[4] The Israeli Green Regulators forum includes the financial regulators in Israel (Chairman of the Capital Market, Insurance and Savings Authority, Chairman of the Securities Authority, and the Supervisor of Banks), and representatives from various government ministries, including the Ministry of Finance, The Ministry of Justice (Economic Department), the National Economic Council and the Ministry of Environmental Protection. The Israel Democracy Institute facilitates the forum and Prof. Karnit Flug, former Governor of the Bank of Israel, moderates the forum.

[5] The Israeli Mitigation Taxonomy draft on MoEP website https://www.gov.il/he/departments/publications/Call_for_bids/taxonomy.

1. Compliance with the technical screening criteria for contributing substantially to the environmental goal. For instance, in the mitigation Taxonomy, this implies meeting the technical screening criteria for a significant contribution to mitigating greenhouse gas emissions.
2. Compliance with the DNSH (Do No Significant Harm) technical screening criteria for the five environmental goals.
3. Meeting the minimal social safeguards as defined in the guidelines of the OECD (2023), the UN (2011), the International Labor Organization (2022), and the International Bill of Human Rights.

The classification, according to the Taxonomy, is binary. In simpler terms, economic activity can align with the Taxonomy if it meets all the relevant technical criteria or not align if it fails to meet one or more of the relevant technical criteria. There are significant differences and adjustments between the European Taxonomy and the Israeli Taxonomy:

1. Changes in the primary sections—for instance, the exclusion of the forestry section from the Israeli Taxonomy was based on the different conditions in the Israeli and European sectors regarding the relevance of ESG standardization, particularly in the context of the capital market. The Jewish National Fund, a non-profit charitable company, manages forests in Israel.
2. Adjustments in economic activities—activities currently irrelevant to the Israeli market, such as nuclear energy, were excluded from the Israeli Taxonomy.
3. Revisions and modifications in the technical criteria cover significant contributions to mitigation and DNSH (Do No Significant Harm) criteria. Each criterion was carefully examined, and if the EU criteria were not applicable due to variances in market conditions or regulatory frameworks, extensive professional effort was invested to develop the most precise and relevant criteria for the Israeli Taxonomy. These differences arise, among other things, from the disparity between the Israeli and the European markets and environmental regulations.

The Israeli Mitigation Taxonomy, voluntarily introduced in early 2024 and modeled after the European Taxonomy, aspires to transform into a compulsory regulatory framework in Israel. The Israeli Ministry of Environmental Protection envisions integrating the Taxonomy into Israeli regulations, compelling market entities to disclose their alignment and consequently improving the data on sustainable investments. Achieving mandatory status in Israel is contingent on providing financial incentives and formulating a regulatory framework through fostering collaboration with financial regulators and the government. It is to be noted that taxonomy classification and reporting is a resource-intensive process for companies that requires learning and specialization.

Nevertheless, companies would benefit from initiating the processes required to generate taxonomy reports as soon as possible. Doing so can enhance their transparency and environmental reputation and attract environmentally conscious investors. Moreover, voluntary reporting will prepare these companies for a future where such reporting may become mandatory. Additionally, there are plans to formulate an

Israeli Taxonomy for the remaining five environmental objectives, with the forthcoming release of the Israeli Adaptation Taxonomy for public consultation. In the future, the goal is to reach a level of adoption and compliance similar to EU standards.

9.3 Sustainable Finance in Practice

We identified seven main dimensions of regulatory implementation of Sustainable Finance in Israel, including the following:

Green Bonds

The Israeli Securities Authority (ISA) introduced guidelines in 2018 to encourage the issuance of green bonds. The guidelines provide a framework for issuing bonds allocated to finance environmentally friendly projects. As of September 2021, there have been several examples of green bond issuances in Israel, aligned with the International Capital Market Association (2021) guidelines, and including:

- Israel Electric Corporation (IEC) Green Bonds: In 2017, the Israel Electric Corporation issued green bonds worth 2 billion NIS (approximately USD 570 million). The proceeds from these bonds were earmarked for financing renewable energy projects, including solar and wind power.
- Bank Hapoalim Green Bonds: In 2019, Bank Hapoalim, one of Israel's largest banks, issued green bonds worth 2 billion NIS (approximately USD 570 million). The funds raised through these bonds were directed to support green projects, such as renewable energy, energy efficiency, and sustainable transportation.
- Meitav Dash Asset Management Green Bonds: Meitav Dash Asset Management, a leading Israeli investment firm, has also been involved in green bond assurances. In 2019, they launched a green bond fund that invests in green bonds issued by Israeli and international entities, supporting various environmentally friendly projects.
- Green Government Bonds (January 2023): Israel's Ministry of Finance successfully issued the country's first-ever Green Government Bonds, amounting to USD 2 billion. These bonds were issued according to the Green Bond Principles set by the ICMA and the technical criteria outlined in Israel's Taxonomy.

Indices

Several indices were integrated into the Israeli Stock Exchange to encourage investments in sustainability. In 2005, the first index published was the MAALA ESG index traded on the Tel Aviv Stock Exchange. In 2020, the TA-Cleantech index and TA-125 Fossil-Fuel-Free Climate Index were introduced.

Enhanced Banking Reporting

The Banking Supervision Department within the Israeli Central Bank has introduced new environmental, social, and governance (ESG) reporting requirements. These amendments are intended to improve banking institutions' public disclosure of environmental and social aspects, mainly focusing on their exposure to climate and other environmental risks and their impacts on society and communities. The Israeli authorities have not mandated specific sustainability reporting requirements for companies. However, the Israeli Institute of Certified Public Accountants (ICPA) issued guidelines in 2020 that encourage companies to disclose ESG-related information in their financial reports voluntarily. The three largest banks in Israel, Poalim, Leumi, and Discount, published voluntary ESG reports.

As for Climate-Related Disclosures, the ISA has suggested that companies provide climate-related disclosures in their financial reports. These disclosures should consider the potential impact of climate change on the company's operations, risks, and opportunities. In practice, only 13% of the publicly traded companies have provided disclosures in 2022.

Commitment to Carbon Neutrality

Israel joined the Green Government's Initiative (GGI) led by the United States, committing to a plan to achieve carbon neutrality in government operations by 2050. This includes the government's direct investment in the transition into green governmental buildings, green transportation, and more. As it relates to the Israeli Nationally Determined Contribution to reducing carbon emissions (Ref: NDC), the Israeli authorities have adopted financial mechanisms supporting cleaner transportation (increasing shares of hybrid and electric cars) and investments in renewable energy resources (including solar infrastructures and wind infrastructures). For example, to stimulate the reduction of carbon emissions in the transportation sector, Israel has rolled out incentives, particularly in the form of reduced taxes on eco-friendly vehicles, including electric and hydro-electric cars. Since 2009, the acquisition tax for electric cars has been set at a mere 10%, which incentivizes their adoption. In contrast, the tax rate for electric-hybrid cars stood at 20%, while hybrid vehicles incurred a 30% tax rate. In stark contrast, traditional fuel-based cars bore a significantly higher tax burden, reaching up to 83%. However, it's worth noting that in 2019, Israel re-calibrated its tax structure, increasing acquisition tax rates for electric-hybrid cars to 25% and further escalating taxes on hybrid vehicles to 45%. This dynamic policy landscape underscores Israel's evolving stance towards sustainable and eco-conscious transportation solutions. This approach reflects a transition from a 'Command and Control' approach, adopted in the 80s, into a market-incentive approach.

9.4 Investment and Risk

The Israeli government has recognized the importance of investor risk in sustainable finance. Efforts have been made to raise awareness among investors about the financial risks and opportunities associated with climate change and sustainability. In June 2023, the banking sector's regulatory authority issued guidelines on "Principles for Effective Climate-Related Financial Risk Management" based on the recommendations of the Basel Committee. These principles enhance the banking sector's preparedness for climate-related financial risks. Moreover, the CMISA has recommended that companies consider ESG factors in their corporate governance practices. This includes ensuring that boards of directors have sufficient expertise in sustainability matters and establishing board-level committees to oversee ESG issues.

Investment Pathways for Sustainability and ESG Integration (September 2022): Israel's Securities Authority published updates to investment pathways for pension savings and long-term savings that facilitate the creation of two new pathways focused on sustainability and the environment. A directive was issued requiring pension and long-term savings fund managers to consider environmental, social, and governance (ESG) factors when determining investment policies. These requirements were published in November 2021 and aim to ensure that ESG considerations are integrated into investment policies. 6 Saving institutions and financial investment companies offer ESG investment, like Israeli pension company Migdal and the private investment firm Value2.

Investment Pathways for Sustainability

In September 2022, Israel's Securities Authority published updates to investment pathways for pension savings and long-term savings that facilitate the creation of two new pathways focused on sustainability and the environment. A directive was issued requiring pension and long-term savings fund managers to consider environmental, social, and governance (ESG) factors when determining investment policies. These requirements were published in November 2021 and aim to ensure that

ESG considerations are integrated into investment policies.6 Saving institutions and financial investment companies offer ESG investment, like Israeli pension company Migdal and the private investment firm Value-2.

Guidance for Funds, Licensed Entities, and Consultancies

The Israel Securities Authority released in December 2022 the final version of guidance on incorporating ESG considerations into investment decision-making and risk management processes for fund managers, large licensed entities, and advisory firms. These adaptations demonstrate Israel's commitment to integrating sustainability and climate-related factors into its financial and governmental operations, aligning with global trends in sustainable finance.

9.5 Looking Forward

In this chapter, we provided a comprehensive overview of the evolution of sustainable finance in Israel, highlighting key regulatory milestones that have significantly shaped the sector. The regulatory journey began in 2009 with the issuance of general instructions aimed at addressing environmental risk hazards in investment corporations and banks. This initial step was followed by the introduction of disclosure regulations in 2012, which enhanced transparency within the financial system. The mandatory adoption of corporate responsibility reports in 2021 further embedded sustainability into corporate practices. The most recent milestone, the adoption of the Israeli taxonomy in 2024, demonstrates Israel's ongoing commitment to aligning with international sustainability standards, particularly the EU's sustainable finance directive approach. As Israel moves forward on its path toward sustainable finance, several key initiatives and trends are expected to shape its financial landscape in the coming years.

1. Regulatory Evolution: Israel's approach to sustainable finance regulation is transforming dynamically. The Israel Securities Authority (ISA) has recently drafted guidelines mandating fund managers to incorporate ESG factors into their investment analysis and decision-making processes. This move signifies Israel's dedication to aligning itself with international best practices and remain competitive in the global financial market. In tandem with these developments, there is an anticipated shift from the current voluntary measures to more stringent, mandatory regulations. Such a transition would mirror the advancements within the European Union, ensuring that the Israeli financial ecosystem adheres to international standards and actively contributes to the global momentum towards sustainable investing.

2. Climate Resilience: Israel's focus on climate stress testing and developing a sustainable risk management framework for banks signifies a growing recognition of the systemic risks posed by climate change. The results of these stress tests are likely to inform future policies and strategies for enhancing climate resilience in the financial sector, potentially leading to new requirements for banks and insurers to assess and disclose their climate-related risks.
3. Green Financing: Israel's foray into issuing green bonds, starting with the Green Bond Framework, is a significant step toward raising capital for sustainable projects and infrastructure. This trend is expected to continue, with the issuance of government green bonds in the pipeline. As green financing gains prominence, it will likely attract domestic and international investors interested in supporting environmentally beneficial initiatives.
4. International Collaboration: Israel's participation in global initiatives, such as the Network of Central Banks for Greening the Financial System (NGFS), underscores its commitment to aligning with international standards. Continued collaboration with global partners will provide Israel with access to best practices, data, and expertise in sustainable finance, helping it stay at the forefront of the rapidly evolving field.
5. Innovation and Technology: Israel's renowned innovation ecosystem is expected to play a pivotal role in advancing sustainable finance. Fintech and StartUps may develop solutions that facilitate ESG reporting, impact measurement, and sustainable investment. These innovations can enhance the efficiency and effectiveness of sustainable finance practices.

In conclusion, Israel's sustainable finance journey is on an upward trajectory, driven by regulatory commitment, market demand, and global imperatives. As the country navigates the challenges and opportunities of sustainable finance, it will contribute to both the resilience of its financial sector and the achievement of broader sustainability goals, ultimately benefiting its economy and society.

References

Bolt, K., Matete, M., & Clemens, M. (2002). *Manual for calculating adjusted net savings*. World Bank Publication. World Bank.

Curtis, M., & Chertoff, M. S. (1973). *Israel: social structure and change. Sociology, political science*. Transaction Books.

European Commission. (2022). EU Taxonomy: Accelerating Sustainable Investments.

European Union Commission Delegated Regulation. (2020a). Regulation 2021/2139. Journal of the European Union, L 198/13. Commission Delegated Regulation (EU) 2021/2139.

European Union Commission Delegated Regulation. (2020b). EU Taxonomy regulation - EU Regulation 2020/852.

International Capital Market Association. (2021). *Green Bond Principles: Voluntary Process Guidelines for Issuing Green Bonds*. https://www.icmagroup.org/green-social-and-sustainability-bonds/green-bond-principles-gbp

International Labor Organization. (2022). X ILO Declaration on Fundamental Principles and Rights at Work and its Follow-up. ILO Declaration on Fundamental Principles and Rights at Work.

Lotan, A., Kost, R., Mandelik, Y., Peled, Y., Chakuki, D., Zemah Shamir, S., & Ram, Y. (2018). National scale mapping of ecosystem services in Israel – genetic resources, pollination, and cultural services. *One Ecosystem, 3*, e25494.

OECD. (2023). *OECD guidelines for multinational enterprises on responsible business conduct*. OECD Publishing. https://doi.org/10.1787/81f92357-en

Sibylle, H. (2010). Communities in processes of change: Comparative differences in entrepreneurial motivation. *International Journal of Innovation and Regional Development, 2*. https://doi.org/10.1504/IJIRD.2010.029853

UN. (2011). Guiding Principles on Business and Human Rights. United Nations Human Rights. UN Guiding Principles on Business and Human Rights.

Vinrab, A., & Shraverman, K. (2023). *Demographic trends in Israel*. Taub Center. https://www.taubcenter.org.il/wp-content/uploads/2023/05/Migration-patterns-HEB-2023.pdf

Dafna M. DiSegni is a Visiting Professor of Economics at the University of Rhode Island. Her research focuses on natural resource markets, industrial organization and the development of financial mechanisms for sustainability. Her research includes includes risk analysis and the valuation of the potential impacts of financial regulations on industry value chains, resource conservation, the development of backstop technologies, and the diffusion of knowledge, information, and sustainability. At the time of ERASMUS project she was affiliated with the Tel Hai College, Israel.

Nadav Paldi is an Environmental Scientist, affiliated with the Israel Ministry of the Environment, contributing to the implementation of Sustainable Finance as it relates to Environmental and Natural Resources Exploitation in the East Mediterranean region.

Gal Tamir Senior Scientist, Digital Data Unit, Government of Israel & Senior Executive with experience in proactive entrepreneurial management, establishing companies and leading them to operation and profitability in Israel and abroad.

Part III
Comparative Insights and Cross-Country Themes in Sustainable Finance

Chapter 10
Social Inclusion and Community Participation in Driving Sustainable Financial Ecosystems

Aleksandra Kekkonen

10.1 Introduction

Social inclusion and community involvement are crucial to achieving the goals of sustainable finance. They ensure that financial services are accessible, effective, and beneficial to all members of society, contributing to overall economic resilience and sustainability. Social inclusion is a process in which all individuals, regardless of background, have the opportunity and resources to participate fully in economic, social, political, and cultural life (UN DESA, 2016). This participation is facilitated by removing institutional barriers and enhancing the capabilities of disadvantaged groups. Social inclusion emphasizes the importance of opportunity, equality, dignity, and respect for all, regardless of race, gender, disability, or socioeconomic status. Community involvement is a process by which community members are brought together to act on issues they care about. It emphasizes participation, collaboration, and shared decision-making (Hasan, 2022). Community involvement can take various forms, from consultation and communication to empowerment, where community members are actively involved in project design, decision-making, and implementation. In the context of development or finance, community involvement often ensures that projects or initiatives align with those directly affected needs, values, and priorities. Sustainable finance and community behavior are deeply interlinked. Sustainable finance relies on community behavior to drive demand and adopt sustainable practices. In contrast, the growth and availability of sustainable finance options can significantly influence and support positive changes in community behavior. Together, they create a powerful synergy for advancing sustainability and addressing critical global challenges.

A. Kekkonen (✉)
Estonian Business School, Tallinn, Estonia
e-mail: Alexandra.kekkonen@ebs.ee

R. Calipha et al. (eds.), *Sustainable Finance Regulation in the European Union*, Sustainable Finance, https://doi.org/10.1007/978-3-031-88392-7_10

Both social inclusion and community involvement are interrelated. Inclusion often requires active community engagement to understand the needs and aspirations of marginalized groups. Similarly, effective community involvement can promote social inclusion by ensuring diverse voices are heard and considered in decision-making processes. Social inclusion and community involvement are critical elements in sustainable finance (Tanrıkul, 2023). The chapter aims to analyze the connection between social inclusion and community involvement, provide an overview of why social inclusion and community involvement are essential for sustainable finance, and present the current body of research in the field. It will discuss examples of community involvement practices and inclusive finance cases of kibbutzim in Israel and Mondragon Corporation in Spain. In the last part of the chapter, the list of challenges to obtaining higher community involvement and policy implications are presented.

10.2 Social Inclusion and Community Involvement as Elements Essential for Sustainable Finance

Financial inclusion focuses on delivering financial services at an affordable cost to all levels of society. It enables people to manage their financial obligations efficiently, reduces poverty, and supports broader economic growth (Demirguc-Kunt et al., 2017). First, it reduces individuals' vulnerability. For instance, facilitating saving allows people to weather shocks and invest in their education, health, and micro-businesses. Second, it increases the efficiency of daily life: bills can be paid electronically without time off work. Third, financial inclusion allows the socialization and diversification of peoples' financial risks through the financial system. For instance, breadwinner insurance can prevent people from falling back into poverty. Fourth, financial inclusion supports economic growth through increasing financial resources to support actual economic activity, particularly for individuals and small and medium enterprises (Zhang-Zhang et al., 2020).

As outlined in the UN SDGs, financial inclusion is crucial to address today's global challenges. Financial access (mainly through FinTech) is one way to reduce the burden of life's challenges, including sickness, crime, poverty, unemployment, age, etc. Financially excluded individuals lack the tools to prepare for and manage such risks (Arner et al., 2020). For instance, farmers without access to electronic payment systems worry about theft and may consume more immediately rather than take the risk. Health insurance can secure one's long-term working capacity. Savings can fund children's education and provide for old age. These are long-term goals.

Financial exclusion allows people to think, plan, and act long-term. Where risks that could be avoided, hedged, or socialized through the financial system materialize, the excluded are forced to think and act short-term, often unsustainably (Arner et al., 2020).

Fintech, or financial technology, is crucial in advancing sustainable finance in several ways. By leveraging innovative technologies, fintech firms can facilitate, enhance, and often democratize access to sustainable financial products and services. Through fintech, sustainable finance becomes more integrated, transparent, and aligned with broader societal goals. Increasing financial inclusion is being seen not as an end but as one fundamental support for achieving broader sustainable development objectives, including the UN SDGs (Arner et al., 2020). In turn, inclusive financial development has a more substantial effect on improving environmental quality (Wang et al., 2022). Improving the convenience and depth of use of financial services is the key to effectively reducing rural household economic vulnerability and improving environmental quality.

There are numerous reasons why social inclusion and community involvement are essential for sustainable finance. Sustainable finance is not just about generating profits but ensuring that the interests of all stakeholders, including marginalized groups, are considered. Promoting social inclusion, financial activities, and services can cater to a broader audience, reflecting diverse needs and interests. Thus, social inclusion and community involvement reflect broader stakeholder interests (Jeffery, 2009). A more inclusive financial system can mitigate economic shocks by diversifying assets and reaching underserved sectors of the economy to reduce systemic risks. Community involvement ensures that the benefits of financial activities, such as infrastructure projects, are fairly distributed. This can prevent wealth concentration and reduce social inequalities.

When the community is involved in decision-making processes, there is a higher likelihood that the financial initiatives will be accepted and trusted. This trust is pivotal for financial projects' long-term success and increasing credibility. Some empirical studies demonstrate the impact of financial inclusion on household economic vulnerability and environmental sustainability. Inclusive credit and insurance services can directly increase rural families' access to inferior rural groups and financial services (Erlando et al., 2020). At the same time, inclusive financial services improve family income by promoting entrepreneurial and innovative activities and alleviating educational poverty (Lensink et al., 2017; Camargo & Stein, 2022). Others propose that the level of coverage of banking institutions at the provincial and community levels can improve credit availability, financial service efficiency, and service quality and significantly reduce household economic vulnerability (Mina Imai, 2017). Inclusive financial development can reduce rural household economic vulnerability and improve environmental quality, particularly in rural areas (Wang et al., 2022).

10.3 Evidence from the Field

Several practical cases will be described to demonstrate how community involvement and social inclusion can be practically implemented: the kibbutz in Israel and the case of the Mondragon cooperative. The author of the chapter was inspired by both examples of participating in field trips as a part of the Erasmus+ IFI project. Implemented elements of community involvement and financial and social inclusion will be analyzed for both cases.

The Kibbutz

The kibbutz is an example of community involvement and social inclusion but with specific characteristics unique to its historical, cultural, and socio-economic context. A kibbutz is a collective community in Israel, traditionally based on agriculture. The first kibbutzim (plural for kibbutz) was founded in the early twentieth century as part of the Zionist movement. Over the years, many kibbutzim have transitioned from purely agricultural communities to ones that also encompass industry and tourism. Community involvement in the kibbutz is implemented through collective decision-making, shared work, and collective upbringing (Amit-Cohen & Sofer, 2016). Kibbutz members traditionally make decisions collectively, usually in general assemblies or through elected committees. This embodies the essence of community involvement, as all members have a say in community affairs. Members of the kibbutz traditionally share work responsibilities. Whether farming, teaching, cooking, or any other task, the workload is distributed among members, fostering a sense of shared purpose and community. In some traditional kibbutzim, children were raised collectively, living together in children's houses rather than with their parents.

This practice emphasized the community's role in child-rearing, although it has become less common in recent decades. Regarding social and financial inclusion, the following aspects are critical when we analyze kibbutz: egalitarian principles, shared resources, and holistic inclusion (Palgi & Getz, 2014). Kibbutzim have traditionally operated on egalitarian principles, where members have equal rights, access to resources, and an equal say in decisions. This promotes social inclusion by ensuring no member is excluded or marginalized based on socio-economic status. Resources, from housing to food to entertainment, have traditionally been shared among kibbutz members. The idea is to ensure everyone can access what they need, reinforcing inclusivity.

Beyond just economic inclusion, the kibbutz model also encompasses social, cultural, and educational inclusion. Activities, cultural events, and educational opportunities are often organized for the benefit of all members. The kibbutz is an example of community involvement and social inclusion within its specific cultural and historical context. However, like all communal experiments, it has faced challenges and has evolved in response to internal and external pressures. The traditional kibbutz model has evolved over the years. Many kibbutzim have undergone significant changes, moving away from strict collectivism and integrating more capitalistic elements (Abramitzky et al., 2022). For instance, some kibbutzim now allow private ownership, pay salaries based on jobs rather than equal distribution, and have diversified their economic activities.

Mondragon Corporation

The Mondragon Corporation, a cooperative federation based in the Basque region of Spain, is an example of financial inclusion and community involvement. Founded in 1956, Mondragon has become one of the world's largest cooperatives, encompassing various sectors from finance to manufacturing to education. Financial inclusion in Mondragon's scope of activities can be illustrated through examples of Caja Laboral, cooperative financial institutions, and risk-sharing mechanisms (Morris, 1992). Caja Laboral, the principal financial institution supporting the Basque cooperative movement, operates as the financial arm of the Mondragon Corporation, delivering banking and financial services to its clientele through a comprehensive network encompassing more than 370 branch offices within and beyond the Basque Country. Comprising over 1800 employees who concurrently serve as partner-owners and aligned with the cooperative ethos of collective ownership and custodianship, Caja Laboral achieves an annual revenue surpassing €330 million. As a cooperative bank, it emphasizes the needs and interests of its members, promoting financial products that benefit its community of users. As a risk-sharing example, the Mondragon framework has a mutual financial support mechanism (Imaz et al., 2023). If a cooperative within the federation faces economic challenges, others can provide financial assistance.

This inter-cooperative solidarity can be viewed as a form of financial inclusion, ensuring that member entities have support in times of need. Community involvement in Mondragon Corporation is implemented by democratic management and participatory decision-making, education and training, local and regional development, and social transformation (Forcadell, 2005; Rodríguez-Oramas et al., 2022). One of the central tenets of Mondragon is "one member, one vote," regardless of the amount of capital each member has invested. This ensures all members have an equal say in the cooperative's decision-making processes. Members participate in decision-making both directly and through elected representatives. This ensures that the perspectives and concerns of the broader membership influence the cooperative's direction. Mondragon places a significant emphasis on the education and training of its members. Through its educational institutions and continuous training programs, it ensures that members are actively involved in shaping and evolving the cooperative's future. The cooperative has consistently prioritized the development of its local and regional community. By creating employment opportunities, investing in local initiatives, and supporting regional growth, Mondragon demonstrates a deep commitment to community involvement. Mondragon's roots lie in the desire for social transformation, with its founder, Father José María Arizmendiarrieta, envisioning the cooperative to address economic challenges in the post-war Basque region. The cooperative's principles emphasize solidarity, cooperation, and placing community and people before capital.

The Mondragon Corporation provides a compelling example of cooperative principles driving financial inclusion and community involvement. Its success showcases the potential of collaborative models to create sustainable and inclusive economic structures that prioritize the well-being of members and the broader community. At the same time, Mondragon Cooperative currently faces many difficulties in implementing financial inclusion and community involvement. As Mondragon grows and globalizes, maintaining a cohesive sense of community involvement becomes challenging (Flecha Ngai, 2014). As cooperatives grow and spread geographically, maintaining direct worker involvement and influence in decision-making can be complex. During economic recessions, tough decisions about layoffs and wage reductions may have to be made. While Mondragon prioritizes job security, downturns can strain the cooperative model. For instance, the bankruptcy of its flagship cooperative, Fagor Electrodoméstіcos, in 2013 highlighted this challenge (Errasti et al., 2017). As Mondragon seeks growth and global competitiveness, it faces challenges maintaining its foundational principles of solidarity, equality, and worker involvement. This balancing act is a perennial challenge.

Mondragon Cooperative and the kibbutz system represent community-oriented models emphasizing collective ownership, democratic decision-making, and shared benefits. While they have emerged from different historical, cultural, and geographic contexts (Mondragon in the Basque region of Spain and the kibbutz system in Israel), they share commonalities in their approaches to financial inclusion and community involvement. Both models are anchored in collective ownership. In Mondragon, workers operate the cooperative, while in traditional kibbutzim, members commonly own all assets and resources. Both systems emphasize democratic governance. In Mondragon, workers have a say in significant decisions, and in kibbutzim, decisions are traditionally made in general assemblies where all members can participate. Both the kibbutz and Mondragon models prioritize community solidarity and cohesion. Their financial and operational structures are designed to promote collective well-being over individual profit. Both Mondragon and the kibbutz system showcase the possibilities of alternative economic and social models that prioritize community involvement and financial inclusion. They are real-world examples of how such models can thrive and adapt amidst changing global economic landscapes.

10.4 Sustainable Finance Barriers and Challenges for Obtaining Higher Community Involvement

'Sustainable finance' and 'community involvement' are two concepts that are intrinsically linked, especially when discussing long-term financial strategies that aim to bring about positive environmental, social, and economic change. As communities become more aware of environmental and social issues, they increasingly demand financial products and services that align with their values. This shift in behavior drives financial institutions to offer sustainable investing options, green bonds, and other ESG-focused products. At the same time, sustainable finance initiatives can focus on projects that bring about immediate and direct benefits to communities.

These can range from creating green jobs and improving infrastructure to promoting health and education (Demirguc-Kunt et al., 2017).

Many sustainable finance efforts emphasize environmental conservation, which directly affects communities. Protecting local environments ensures the sustainability of resources that communities rely on, like clean water, arable land, and breathable air (Wang et al., 2022, p. 4). Sustainable finance promotes inclusive financial practices, ensuring marginalized communities can access banking, loans, and other financial products. This can spur local entrepreneurship, improve access to essential services, and reduce economic disparities. A foundational principle of sustainable finance is engaging various stakeholders, including local communities. This ensures that projects are tailored to community needs, resulting in a more significant and lasting positive impact. While aiming to achieve long-term environmental, social, and economic benefits, sustainable finance encounters barriers and challenges when trying to increase community involvement. Below there is a list of challenges to obtaining higher community involvement (Zhang-Zhang et al., 2020; Arner et al., 2020):

- Many communities may not be familiar with sustainable finance or its potential benefits. Without understanding the value proposition, it is challenging to get communities involved.
- Sustainable finance initiatives may only sometimes resonate with local cultures or address communities' immediate socio-economic challenges. For instance, a community facing immediate economic hardships might prioritize short-term gains over longer-term sustainable objectives.
- Past negative experiences with financial institutions or poorly executed sustainable projects can lead to mistrust. Communities might be wary of external financial actors and their intentions.
- Limited access to financial services exists in many regions, mainly rural areas. This infrastructural challenge can hinder community involvement in sustainable finance initiatives.
- Communities might lack the skills or knowledge to participate in or drive sustainable finance projects actively. This capacity gap can limit their ability to engage meaningfully.

Policymakers, financial institutions, and community leaders must work to overcome these barriers. Strategies might include capacity-building, transparent communication, inclusive decision-making, and creating regulatory environments prioritizing community-driven sustainable finance.

10.5 Policy Implications

Policies are crucial in shaping the financial landscape and determining how inclusive it is. When designed with social inclusion and community involvement in mind, policies can significantly enhance the ability of sustainable finance initiatives to deliver broader and more lasting impacts (Littig & Griessler, 2005). Policies can prioritize social inclusion and community involvement by setting them as central tenets of sustainable finance frameworks. Such policies can ensure that financial practices actively consider marginalized groups and local community needs. At the same time, if policies overlook or marginalize these issues, sustainable finance may focus solely on environmental or direct economic outcomes, leaving behind significant portions of the population. Policies can establish regulatory frameworks that incentivize financial institutions to integrate social inclusion criteria into their lending, investing, and operational practices. However, weak regulations or policies emphasizing profit without social considerations can lead to exploitative practices and hinder genuine community involvement. Policies can offer tax breaks, subsidies, or grants to institutions prioritizing socially inclusive sustainable finance initiatives (UN, 2023).

However, misaligned financial incentives might prioritize specific industries or big corporations over local community needs or exclude marginalized groups. Policies can establish formal community consultations and feedback mechanisms, ensuring financial decisions reflect local needs and priorities. However, the absence of such mechanisms can lead to top-down decision making, alienating communities and potentially leading to unsustainable or detrimental projects. Policies can institute quotas or affirmative actions to ensure the representation of marginalized groups in decision-making bodies related to sustainable finance.

If not carefully implemented, these can be viewed as tokenistic or result in backlash (Mirzoev et al., 2022). Policies can fund infrastructure or technology projects that enable broader community participation in sustainable finance, such as mobile banking solutions in remote areas (Arner et al., 2020, p. 18). Failure to invest in such infrastructures can further exclude marginalized communities from the financial system. Policies are instrumental levers in shaping the role of social inclusion and community involvement in sustainable finance. While the right policies can catalyze inclusive and community centered finance initiatives, misaligned policies can exacerbate inequalities or dilute the essence of sustainability. Decision-makers must thus be aware of the profound impacts their policy choices can have in this domain.

10.6 Conclusions

Sustainable finance and community involvement are deeply interconnected. Financial systems prioritizing sustainability must inherently value social inclusion and community participation to address socioeconomic disparities and ensure long-term holistic growth. The field of sustainable finance is evolving, with new challenges and opportunities emerging continually. Community involvement ensures adaptability and resilience amidst these changes. Financial institutions should prioritize and invest in platforms that allow for robust community engagement, ensuring that diverse voices are heard and integrated into decision-making processes.

Governments and regulatory bodies should create enabling environments for sustainable finance models prioritizing community involvement. This could be tax incentives, grants, or favorable lending rates for community-focused projects. Regular reviews are essential given the dynamic nature of communities and financial systems. Feedback mechanisms should be implemented to ensure systems remain relevant and practical. For community trust and genuine involvement, transparency in financial operations, decision-making processes, and outcomes is crucial.

References

Abramitzky, R., Ben-Porath, N., Lahad, S., Lavy, V., & Palgi, M. (2022). *The effect of labour market liberalization on political behavior and free market norms*. Working Paper No. 30186. National Bureau of Economic Research. http://www.nber.org/papers/w30186

Amit-Cohen, I., & Sofer, M. (2016). Cultural heritage and its economic potential in rural society: The case of the kibbutzim in Israel. *Land Use Policy, 57*, 368–376. https://doi.org/10.1016/j.landusepol.2016.05.031

Arner, D. W., Buckley, R. P., Zetzsche, D. A., & Veidt, R. (2020). Sustainability, FinTech and financial inclusion. *European Business Organization Law Review, 21*, 7–35. https://ssrn.com/abstract=3387359

Camargo, B., & Stein, G. (2022). Credit constraints and human capital policies. *Journal of Public Economics, 208*(C). https://doi.org/10.1016/j.jpubeco.2022.104624

Demirguc-Kunt, A., Klapper, L., & Singer, D. (2017). *Financial inclusion and inclusive growth: A review of recent empirical evidence*. Policy Research Working Paper No. 8040. Development Research Group, Finance and Private Sector Development Team. https://documents1.worldbank.org/curated/en/403611493134249446/pdf/WPS8040.pdf

Erlando, A., Riyanto, F. D., & Masakazu, S. (2020). Financial inclusion, economic growth, and poverty alleviation: Evidence from eastern Indonesia. *Heliyon, 6*(10), e05235. https://doi.org/10.1016/j.heliyon.2020.e05235

Errasti, A., Bretos, I., & Nunez, A. (2017). The viability of cooperatives: The fall of the Mondragon Cooperative Fagor. *Review of Radical Political Economics, 49*(3), 048661341666653. https://doi.org/10.1177/0486613416666533

Flecha, R., & Ngai, P. (2014). The challenge for Mondragon: Searching for the cooperative values in times of internationalization. *Organization, 21*(5), 666–682. https://doi.org/10.1177/1350508414537625

Forcadell, F. J. (2005). Democracy, cooperation and business success: The case of Mondragón Corporación Cooperativa. *Journal of Business Ethics, 56*(3), 255–274. http://www.jstor.org/stable/25123430

Hasan, M. (2022). *Community development practice: From Canadian and global perspectives*. Centennial College. https://ecampusontario.pressbooks.pub/communitydevelopmentpractice

Imaz, O., Freundlich, F., & Kanpandegi, A. (2023). The governance of multi stakeholder cooperatives in Mondragon: The evolving relationship among purpose, structure, and process. In S. Novković, K. Miner, & C. McMahon (Eds.), *Humanistic governance in democratic organizations (Humanism in Business Series)*. Palgrave Macmillan. https://doi.org/10.1007/978-3-031-17403-2_10

Jeffery, N. (2009). *Stakeholder engagement: A road map to meaningful engagement. How to do Corporate Responsibility' Series*. Doughty Centre, Cranfield School of Management. https://www.fundacionseres.org/lists/informes/attachments/1118/stakeholder%20engagement.pdf

Lensink, R., Servin, R., & van den Berg, M. (2017). Do savings and credit institutions reduce vulnerability? New evidence from Mexico. *Review of Income and Wealth, 63*, 335–352. https://doi.org/10.1111/roiw.12213

Littig, B., & Griessler, E. (2005). Social sustainability: A catchword between political pragmatism and social theory. *International Journal for Sustainable Development, 8*(1/2), 65–79. https://www.ihs.ac.at/fileadmin/public/soziologie/test2.pdf

Mina, C. D., & Imai, K. S. (2017). Estimation of vulnerability to poverty using a multilevel longitudinal model: Evidence from the Philippines. *Journal of Development Studies, 53*(12), 2118–2144. https://doi.org/10.1080/00220388.2016.1265942

Mirzoev, T., Tull, K. I., Winn, N., Mir, G., King, N. V., Wright, J. M., & Gong, Y. Y. (2022). Systematic review of the role of social inclusion within sustainable urban developments. *International Journal of Sustainable Development and World Ecology, 29*(1), 3–17. https://doi.org/10.1080/13504509.2021.1918793

Morris, D. (1992). *The Mondragon system: Cooperation at work*. Institute for Local Self-Reliance and Infinity. https://www.ilsr.org/wp-content/uploads/files/images/mondragon.pdf

Palgi, M. L., & Getz, S. (2014). Varieties in developing sustainability: The case of the Israeli kibbutz. *International Review of Sociology, 24*(1), 38–47. https://doi.org/10.1080/03906701.2014.894344

Rodríguez-Oramas, A., Burgues-Freitas, A., Joanpere, M., & Flecha, R. (2022). Participation and organizational commitment in the Mondragon Group. *Frontiers in Psychology, 13*. https://doi.org/10.3389/fpsyg.2022.806442

Tanrıkul, A. (2023). The role of community participation and social inclusion in successful historic city center regeneration in the Mediterranean region. *Sustainability, 15*(9), 7723. https://doi.org/10.3390/su15097723

United Nations, Department of Economic and Social Affairs. (2016). *Leaving no one behind: The imperative of inclusive development*. Report on the World Social Situation 2016 (ST/ESA/362). New York. https://www.un.org/esa/socdev/rwss/2016/chapter1.pdf

United Nations, Inter-agency Task Force on Financing for Development. (2023). *Financing for sustainable development report 2023: Financing sustainable transformations*. United Nations. https://desapublications.un.org/publications/financing-sustainable-development-report-2023

Wang, Y., Fahad, S., Wei, L., Luo, B., & Luo, J. (2022). Assessing financial development and financial inclusion's role in enhancing environmental sustainability: Do financial inclusion and eco-innovation promote sustainable development? *Frontiers in Environmental Science*. https://doi.org/10.3389/fenvs.2022.1056478

Zhang-Zhang, Y., Rohlfer, S., & Rajasekera, J. (2020). An eco-systematic view of cross-sector fintech: The case of Alibaba and Tencent. *Sustainability, 12*(21), 8907. https://doi.org/10.3390/su12218907

Aleksandra Kekkonen is a Senior Researcher Fellow at the Research, Development & Innovation Unit, Estonian Business School, Estonia.

Chapter 11
Sustainable Finance Implementation: Cross Country Comparison and Prospects

Dafna M. DiSegni

11.1 Comparison of Sustainable Finance Approach

As the European Union advances its Sustainable Finance Regulations, each member state tailors its approach to align with both national priorities and overarching EU directives. This comparative analysis explores how various countries, including Germany, the Netherlands, Italy, Estonia, Spain and Israel, have uniquely adopted and implemented these regulations to foster sustainable finance. The Country-Specific Pathways to Sustainable Finance Implementation is detailed in Part II (Chap. 4, 5, 6, 7, 8, and 9) and key milestones are presented in Figs. 11.1, 11.2, 11.3, 11.4, 11.5 and 11.6. It is evident that the most significant progress in the adoption of sustainable finance practices across countries began in 2020, marked by the implementation of the EU Taxonomy Regulation 2020/852 (EUR-Lex, 2020). This implementation was followed by a gradual development of local regulations focusing on reporting and disclosure. Prior to 2020, there was a decade of convergence during which countries adopted measures such as auditing, ex-ante supervision, and increased involvement of financial institutions in risk management and disclosure practices.

By examining the distinct strategies ranging from technological innovation to financial market adaptations, we gain insights into the diverse yet unified efforts within the EU to promote environmental sustainability and economic resilience across the continent.

The Netherlands and Germany both exhibit strong commitments to sustainable finance, each taking distinct yet complementary approaches within the European framework. The Netherlands is celebrated for its innovative and integrated strategies in sustainable finance, heavily focusing on pioneering financial technologies

D. M. DiSegni (✉)
University of Rhode Island, Kingston, RI, USA
e-mail: dafna.disegni@uri.edu

R. Calipha et al. (eds.), *Sustainable Finance Regulation in the European Union*, Sustainable Finance, https://doi.org/10.1007/978-3-031-88392-7_11

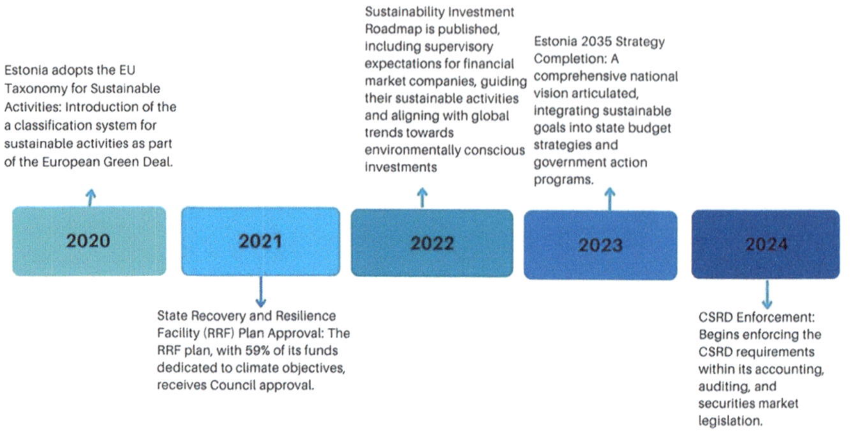

Fig. 11.1 Estonia's development path in sustainable finance implementation

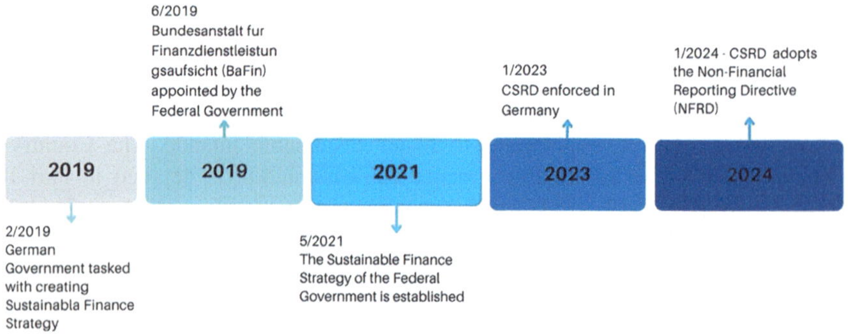

Fig. 11.2 Germany's development path in sustainable finance implementation

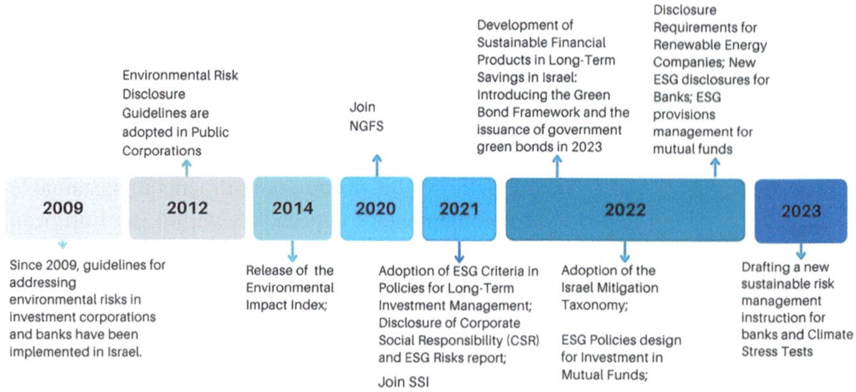

Fig. 11.3 Israel's development path in sustainable finance implementation

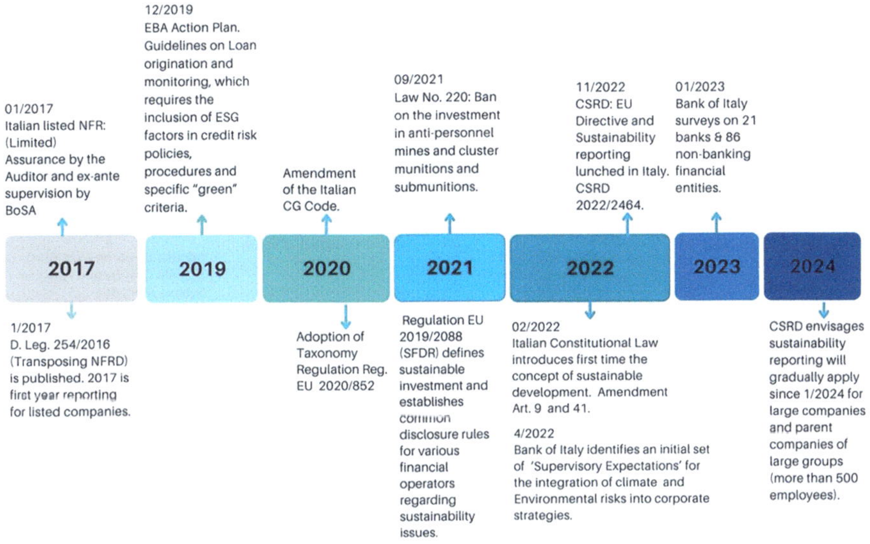

Fig. 11.4 Italy's development path in sustainable finance implementation

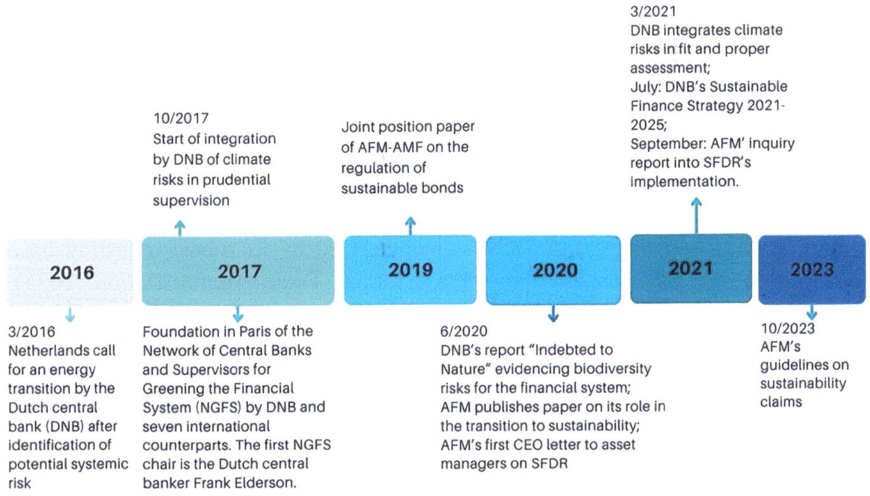

Fig. 11.5 Netherland's development path in sustainable finance implementation

and green investments that support the European Green Deal's goals (Schoenmaker & Véron, 2021). Dutch policies strongly promote environmental sustainability through substantial investments in renewable energy and the adoption of ESG criteria across its financial sector. The Netherlands' approach to sustainable finance is highly proactive and integrated, aligning seamlessly with the European Commission's sustainability objectives. Renowned for its innovative financial landscape and strong

Fig. 11.6 Spain's development path in sustainable finance implementation

emphasis on environmental conservation, the Netherlands actively promotes the adoption of sustainable practices within its financial sector. This includes substantial investments in green technologies and infrastructures, such as renewable energy and sustainable agriculture, which directly support the European Green Deal's ambition to transform the EU into a modern, resource-efficient economy. Dutch financial institutions are at the forefront of incorporating ESG factors into their investment decisions, encouraged by supportive regulatory measures and a national culture that values sustainability. The country's commitment to these principles enhances the European Union's collective efforts in climate change mitigation and fosters a resilient, sustainable financial system across the region. This strategic alignment demonstrates the Netherlands' role as a key player in advancing the European sustainable finance agenda (see Chap. 8).

In contrast, Germany's approach is characterized by its robust industrial base transitioning towards sustainability (International Trade Administration, 2023). German policy initiatives emphasize technological innovation within traditional industries, such as automotive and manufacturing, to reduce environmental impacts. This is complemented by significant investments in renewable energy and the implementation of stringent regulations that ensure financial practices align with sustainability goals. Rooted in the country's renowned technological and engineering excellence, the German model leverages public policy to foster private sector investment in sustainable projects. The approach is underpinned by a strong regulatory framework that includes clear guidelines for environmental, social, and governance (ESG) criteria, alongside incentives for green technology and renewable energy investment. Germany has also focused on integrating sustainability into its financial services sector, encouraging banks and investors to incorporate ESG factors into their risk assessments and investment decisions. The country has established green finance clusters to promote knowledge exchange and innovation, and has actively engaged in international collaborations to align its practices with global sustainability standards. With a proactive government role, comprehensive legislation, and collaborative industry initiatives, Germany's implementation of sustainable finance is both robust and forward-looking. It aims not just to comply with EU mandates but to lead by example, setting high standards for sustainability within the financial sector and across the broader economy. Together, the Netherlands and Germany exemplify the multifaceted nature of sustainable finance within the EU. While the Netherlands leads in financial innovation and sector-wide ESG integration, Germany leverages its industrial strength to foster green technologies and sustainable industrial practices. Both nations contribute powerfully to the broader EU mission of climate change mitigation and economic sustainability, showcasing diverse yet unified approaches to meeting Europe's ambitious sustainability targets.

Italy's approach to sustainable finance implementation is deeply integrated with its cultural emphasis on social responsibility and community engagement. Italian regulators and financial institutions prioritize inclusive finance and community investment, ensuring that sustainability initiatives support local development and contribute to social equity. Italy has been proactive in aligning its national regulations with EU directives, while also fostering public-private partnerships that drive investment in sustainable infrastructure and technologies. Italian banks and investors are increasingly incorporating ESG considerations into their strategies, with a focus on long-term value creation that benefits both the environment and society (Coppotelli, 2022). This approach reflects Italy's commitment to a sustainable finance ecosystem that harmonizes the nation's rich cultural heritage with modern economic resilience (see Chap. 5).

Contrasting with Italy's socially oriented sustainable finance, Spain's approach is significantly shaped by its environmental imperatives, notably its vulnerability to climate change impacts like rising sea levels and droughts. Spanish strategies prioritize investments in renewable energy and infrastructure resilience, reflecting urgent ecological concerns. While Italy emphasizes community and inclusivity, Spain's model leans heavily on technological and environmental innovation to drive its sustainability agenda. Spanish financial institutions and policymakers are focused on developing green financial products and markets, aligning closely with the EU's taxonomy for sustainable activities. This has led to Spain adopting a more targeted approach towards environmental sustainability, making it a central pillar in its overall strategy for sustainable finance, which is distinct from Italy's broader integration of social and cultural considerations (see Chap. 4).

Estonia's approach to sustainable finance is notably influenced by its unique environmental landscape and its strong agricultural traditions. Leveraging its digital leadership, Estonia integrates cutting-edge technologies to enhance the sustainability of its farming practices and to protect its diverse ecosystems. The country emphasizes green innovation in agriculture, aiming to boost productivity while conserving natural resources (KPMG Baltics, Eversheds Sutherland Bitāns (2023). Financial incentives are targeted at modernizing agricultural techniques, reducing emissions, and fostering organic farming. Estonia's sustainable finance framework is also tailored to support initiatives that address the conservation of its vast forests, wetlands, and biodiversity, reflecting its commitment to environmental stewardship. This dual focus on the advancement of sustainable agriculture and the preservation of natural habitats places Estonia at the intersection of traditional industry and ecological preservation within the realm of sustainable finance (see Chap. 7).

Israel's approach to sustainable finance is intricately tied to its unique socioeconomic landscape, defined by limited natural resources and a history of communal living through the kibbutz system. Faced with scarce land and water resources, Israel has prioritized innovation in areas like water technology and desert agriculture, ensuring that sustainability efforts support resource efficiency and innovation. The financial sector reflects this by channeling investments into technologies that maximize resource utilization and foster sustainable practices across industries (see Chap. 9).

In contrast to Estonia's emphasis on leveraging its digital prowess to advance its agricultural sector and environmental conservation, Israel focuses on innovative solutions to overcome its resource limitations. Its community-centric living models, such as the kibbutzim, have also inspired collaborative approaches to sustainable development, emphasizing communal welfare and shared resource management. The kibbutzim's collective ethos is mirrored in Israel's sustainable finance strategies, which prioritize projects that serve broader community interests and foster social cohesion alongside environmental sustainability. This approach not only addresses the country's resource challenges but also reinforces the societal values embedded in its culture.

11.2 Enforcement and Compliance Challenges

The implementation of sustainable finance regulations within the European Union places a significant emphasis on transparency reporting and enforcement. However, this focus presents several challenges that can impact the effectiveness and uniformity of the sustainable finance framework across member states.

Transparency Reporting Challenges

As the European Union intensifies its efforts to integrate sustainability across its financial sectors, transparency in reporting emerges as a fundamental pillar. Yet, achieving comprehensive and consistent transparency in sustainable finance reporting presents a multitude of challenges. These challenges not only stem from the complexity of the data involved but also from varying capabilities and interpretations among financial institutions across member states. Effective solutions to these challenges are critical for ensuring that sustainability reporting is accurate, comparable, and actionable. Here are some key challenges in transparency reporting that need to be addressed to advance the EU's sustainable finance goals.

1. Complexity of Data: The requirement for detailed reporting on environmental, social, and governance (ESG) factors introduces complexities in data collection, verification, and disclosure. Financial institutions may struggle with the granularity of data required, leading to inconsistencies in reporting standards and difficulties in comparing data across entities.
2. Technological Limitations: adequate systems must be in place to handle the large volumes of data required for ESG reporting. Many institutions may lack the technological infrastructure to efficiently gather, process, and report this data, leading to potential delays and inaccuracies.
3. Diverse Interpretation of Standards: different interpretations of what constitutes sustainable activities, particularly under the EU Taxonomy, can lead to varied reporting practices. This divergence complicates the goal of achieving standardized, comparable reports across the EU.

Enforcement Challenges

As the European Union pushes forward with its ambitious sustainable finance agenda, the effective implementation of transparency reporting and robust enforcement mechanisms are crucial. However, these processes are fraught with challenges that can hinder the progress toward achieving uniformity and efficacy across the EU's financial system. The complexity of sustainable finance demands precise and standardized reporting, backed by diligent enforcement to ensure compliance and promote true sustainability in financial practices. Addressing these challenges is essential for maintaining the integrity and success of the EU's sustainable finance initiatives. Let's explore some of the primary enforcement challenges that need to be navigated.

1. Regulatory Alignment: Ensuring that national regulatory bodies are aligned in their interpretation and enforcement of EU sustainable finance regulations is crucial. Discrepancies in enforcement across countries can undermine the integrity of the EU's sustainable finance objectives and create competitive disadvantages.
2. Resource Allocation: Effective enforcement requires that national authorities have sufficient resources, including skilled personnel who understand the nuances of sustainable finance. Many regulatory bodies may face resource constraints that hinder their enforcement capabilities.
3. Cross-Border Coordination: The global nature of financial markets and the operations of multinational corporations complicate enforcement. Coordination between regulatory authorities across different jurisdictions is essential to manage cross-border challenges and ensure comprehensive oversight.
4. Compliance Motivation: While penalties for non-compliance can drive adherence to reporting standards, fostering a culture of transparency and sustainability within organizations is equally important. Ensuring that companies not only

comply out of obligation but also recognize the intrinsic value of sustainability reporting is a challenge.

11.3 Concluding Remarks and Strategic Recommendations

To address these challenges, the EU could consider several strategic actions:

1. Strengthening Guidelines: Clarifying and, if necessary, standardizing reporting guidelines to reduce ambiguity and ensure uniformity in data reporting.
2. Investing in Technology: Supporting financial institutions in upgrading their technological capabilities to handle ESG data efficiently.
3. Enhanced Training: Offering training programs for regulators and financial entities to deepen their understanding of sustainable finance principles and reporting requirements.
4. Cross-Border Regulatory Frameworks: Establishing stronger frameworks for cooperation among EU member states to ensure consistent enforcement across borders.

By addressing these transparency reporting and enforcement challenges, the EU can enhance the robustness and reliability of its sustainable finance framework, thereby supporting its broader objectives of environmental sustainability and social equity. As we conclude this exploration of transparency and reporting challenges in sustainable finance across six diverse EU countries, it becomes clear that while each faces unique hurdles, there are common threads that bind their experiences. These challenges underscore the importance of a unified approach to addressing sustainability within the financial sector, ensuring that standards and practices not only align with the overarching goals of the EU but also adapt flexibly to the specific contexts of individual member states. This book has illuminated the intricate dance between national interests and collective European directives, offering insights into the evolving landscape of sustainable finance. Moving forward, it is imperative for continued collaboration and innovation across borders, driving forward the agenda of sustainability with resilience, accountability, and transparency at its core. As we look to the future, the lessons gleaned from these countries will undoubtedly shape the path towards a more sustainable and financially robust European Union.

References

Coppotelli, C. (March, 2022). "La strategia europea sullo sviluppo sostenibile. In particolare, la finanza sostenibile e le modifiche al quadro regolamentare europeo.", il Mulino.

EUR-Lex. (2020). Regulation (EU) 2020/852 of the European Parliament and of the Council of 18 June 2020 on the establishment of a framework to facilitate sustainable investment, and amending Regulation (EU) 2019/2088. https://eur-lex.europa.eu/legal-content/EN/TXT/?uri=celex%3A32020R0852

International Trade Administration. (2023). *Germany Country Commercial Guide*. https://www.trade.gov/country-commercial-guides/germany-market-overview?section-nav=1956

KPMG Baltics, Eversheds Sutherland Bitāns. (2023). *The Sustainable Finance Roadmap for Estonia*. https://www.kik.ee/sites/default/files/sustainable-finance-roadmap-estonia-2023_en_17.08.pdf

Schoenmaker, D., & Véron, N. (2021). A twin peaks vision for Europe. In A. Godwin & A. Schmulow (Eds.), *The Cambridge handbook of twin peaks financial regulation* (pp. 282–291). Cambridge University Press.

Dafna M. DiSegni is a Visiting Professor of Economics at the University of Rhode Island. Her research focuses on natural resource markets, industrial organization and the development of financial mechanisms for sustainability. Her research includes includes risk analysis and the valuation of the potential impacts of financial regulations on industry value chains, resource conservation, the development of backstop technologies, and the diffusion of knowledge, information, and sustainability. At the time of ERASMUS project she was affiliated with the Tel Hai College, Israel.